Ultimate Guide To Marijuana:

2 Books In 1 - Growing Marijuana & Cannabis & Weed 101 - A Guide To Horticulture, Stocks, Business, Investing, Addiction, Medical Use, Chronic Pain, CBD & Spirituality

Table Of Contents

Introduction 2

Chapter One: Marijuana and Cannabis: History and Basic Knowledge 3

Chapter Two: Seeds: What to Know 8

Chapter 3: Cultivating the Marijuana Plant 12

Chapter Four: Indoor Growing: Setting Up the Indoor Garden 18

Chapter 5: Indoor Growing: Planting and Planting Techniques 23

Chapter 6: Indoor growing. Nutrients and Environmental Factors . 32

Chapter 7: Indoor Growing: Flowering and Breeding 38

Chapter Eight: Indoor Growing: Maintaining the Marijuana Plant .. 44

Chapter Nine: Outdoor Growing: Where and How? 48

Chapter Ten: Outdoor Growing: Planting and Planting Techniques 53

Chapter Eleven: Outdoor Growing: Care of the Growing Plants 58

Chapter Twelve: Outdoor Growing: Nutrients and Environmental Factors 64

Chapter Thirteen: Outdoor Growing: Flowering and Breeding 69

Chapters Fourteen: Harvesting the Marijuana Plants 74

Chapter Fifteen: After Harvesting, What's Next?! 78

Conclusion 81

Introduction	84
Chapter One: Historical Background of Marijuana	86
Chapter Two: The Face of Marijuana Today	93
Chapter Three: Unveiling the Truth Behind Cannabis Addiction	100
Chapter Four: Marijuana Horticulture	106
Chapter Five: Indoor and Outdoor Growing	112
Chapter Six: Sowing Phase, Vegetative Growth Phase, and Flowering Phase	119
Chapter Seven: Harvest of the Cannabis Plant	126
Chapter Eight: Marijuana Growing as a Business Venture	134
Chapter Nine: Steps to take in Starting marijuana Business as a Novice	142
Chapter Ten: Investing in a Marijuana Growing Venture	150
Chapter Eleven: Medical Marijuana (Traditional)	159
Chapter Twelve: Medical Marijuana (Modern)	168
Chapter Thirteen: Marijuana (As a Pain Reliever)	178
Chapter Fourteen: Marijuana and Spirituality	187
Chapter Fifteen: Now You Know Better!	198
Conclusion	202

The Growing Marijuana Handbook

How To Easily Grow Marijuana and Cannabis Indoor and Outdoors Including Tips on Horticulture, Growing in Small Places and Medical Marijuana - For Beginners and Advanced Growers.

Introduction

Today, the name "Cannabis" rings a bell in the minds of every teenager in the world. This is clearly not for its medicinal usage or benefit, but for the reputation of its abuse and irresponsible usage by those who want to "get high". This notion has tainted the cannabis plant to be a deadly poison one must abstain from.

However, this is not true. Cannabis has, in fact, become more of a help than hindrance to man. This concept and more are what this handbook will delve into as it enlightens you even further. Growing cannabis could prove to be a big endeavor at first, but with the help of this handbook, you will easily scale through the steps.

From the beginning to the end, this handbook will cover every detail of marijuana growing. You'll become conversant with not just the life cycle of the cannabis plant but also its uses, benefits, and economic importance. Rest assured, at the end of this handbook comes enlightenment, fresh ideas, and an awesome experience. Unless your plan to grow cannabis is not curtailed by several factors which include the law, financial capability, and time, this handbook will equip you with all the knowledge you need to be an effective grower.

Have you been experiencing bad growth with your cannabis plants? Then this handbook is just perfect for you. It will pinpoint the mistakes you've made so you can correct them next time. It's going to be a bumper harvest - year in, year out. Read through the chapters from beginning to the end, so as to get an idea of what the book is talking about.

In the end, we greatly hope that this handbook will help you achieve your goal of not only growing cannabis, but also growing a healthy-looking cannabis plant.

Sit tight, have fun, and enjoy the ride!

Chapter One

Marijuana and Cannabis: History and Basic Knowledge

Historically, cannabis, which is also known as marijuana, weed, and dope, has existed since the beginning of mankind. It has served as a help to man for both medical and recreational purposes. As a psychoactive herb, the cannabis plant blossoms in an environment which is not devoid of adequate water, sunlight, and temperature.

As one hell of a stubborn plant, cannabis can sprout out of any environment whatsoever – valley, plain, hill or even edgy lands. History has it that the cannabis seeds were mostly planted accidentally around the world either by the winds, birds in the sky, or attached to the hoofs of animals traveling on foot. Little wonder why the cannabis plant is very adaptive to both indoor and outdoor conditions.

Notwithstanding, the historical background and origin of the cannabis plant can be debatable, as there is no consensus on how its life began. Since, it has gained popularity with special improvements, modifications, and upgrades to aid its growth.

When the name "Cannabis" or "Marijuana" pops up, an average individual with little or no knowledge about the plant would most definitely think about the abuse and illegal use of the - which to a large extent has tainted the good name of the cannabis plant. However, there is more to it. The cannabis plant can serve many benefits.

To begin with, cannabis dates back to the era of our ancestors. They used the cannabis plant for quite a number of things. The early 1900s showcased a time when the cannabis plant was vehemently used in the textile industry to create fabrics as a result of its strong nature and ability to withstand a large amount of stress.

This was known as "Hemp" and was proven to be a far better option than the ordinary cotton used today. In reality, what this means is that people had to own minimal clothing items as they could stand the test of time. And guess what? It didn't go down well with the fabric industry. Owning few items means low patronage, and low patronage means low profit.

Now guess what they did next? Your guess is as good as mine! Propaganda and stereotypes were churned up about the cannabis plant during that period. And, yes - they succeeded. The government restricted cannabis to a very large extent. Ever since then, there have been a great number of crusades and campaigns against the cannabis plant.

The stereotypes that go along with the cannabis plant have tainted the real usefulness of the plant, thereby clouding the minds of everyone with propaganda and false ideas. Here are some of the accusations levied on the cannabis plant:

1. Cannabis is abused and harmful.
2. Cannabis is addictive.
3. People die from the intake of cannabis.
4. Cannabis addiction cannot be easily curbed.

These notions are not entirely true. Yes, the cannabis plant is being abused as its leaves are sold as dope, weed, or greens by hoodlums and miscreants. But make no mistake, the decision to abuse the effects of the cannabis plant lies entirely with the abuser.

As a psychoactive herb or plant, the cannabis plant can be very much addictive. According to propaganda, the plant, leaves, and seeds contain the THC, or the tetrahydrocannabinol, which makes the cannabis addictive and hard to rehabilitate. This is not true. Although, THC, or the tetrahydrocannabinol chemical, in the cannabis plant goes a long way in creating hallucinations, delusions, and change of thinking, its addiction lies solely with the abuser.

As a matter of fact, there has been no record of anyone that has died from the intake of cannabis. This notion is simply exaggerated. It is so obvious that death scares would work magic as people tend to stay clear of such activities. Law enforcement agencies have been created, promotions and campaigns embarked on, and laws formulated – thereby hiding the real objective behind these flimsy excuses.

It is important to note that cannabis addiction can be curbed via psychotherapy.

Therefore, when people hear the word "cannabis", what comes to mind is the normal greens, weed, skunk, or dope wrapped in a dirty sheet of paper sold on the street. Far from this, there are lots of advantages and benefits attached to the use of the cannabis plant. This handbook promises to enlighten and hopefully change your perception of the cannabis plant.

Importance of the Cannabis Plant

The importance of the cannabis plant to man can be classified into different segments – Health benefits and Economic benefits.

- **Health Benefits:**
To begin with, research has shown that the cannabis plant has more than 20 health benefits to mankind as a whole. A study on cannabis embarked on by the National Eye Institute in the 1970s shows that the plant can be used to cure Glaucoma (a severe eye defect). It slows down the rapid growth of the disease and, as a result, prevents blindness.

Cannabis also helps to control seizures that occur in people suffering from Epilepsy. This was tested and proven in a study in 2003 by Robert J. Delorenzo and has proven to be effective ever since.

Cannabis also goes a long way in improving the condition of the lungs - unlike its counterpart, tobacco, which damages them. The Journal of the American Medical Association laid emphasis on this.

More than ever today, cannabis intake has proven to be effective in the fight against cancer. The California Pacific Medical Center showcased this in their reports, pointing out how cannabis can stop the spread of cancer further in the host body. What cannabis does is to mute the Id-1 gene, thereby replicating the gene into millions of replicas to fight the cancer by spreading through the host body. This way, cancer is curtailed to an extent.

Have you ever wondered why cannabis users are mostly slim and fit? Well, cannabis intake helps a lot in burning down fat in the body. Not only that - it also helps you achieve a much healthier metabolism.

- **Economic Benefits:**

To a very large extent, cannabis would have a positive impact on the revenue of a government, firm, or even individuals. The sale of this plant, its roots, or its leaves could generate more than enough income, which in turn could raise the standard of living in the region.

Mass growth of the cannabis plant would create what is calculated to be millions of new jobs. Are you wondering where the jobs would come from? It's pretty simple. Jobs like cannabis growers, processors, distributors, retailers, inspectors, and much more along the cannabis line would be available for people to engage themselves in - and in turn, make a living from.

Species

There are basically three types of Cannabis species.

1. **Sativa:** Sativas are tall, lengthy cannabis plants with pointed ends on their leaves. The leaves are also free from any marks or patterns. There is a presence of long internodes which exist in between branches from 3 inches to 6 inches.
2. **Indica:** Indicas are quite short, unlike their tall sativa counterparts. They are the direct opposite of sativa as they possess short internodes and bear marble-like patterns in their round shaped leaves.
3. **Ruderalis:** Ruderalis are also short cannabis plants of about 6 inches long. Their leaves are small and thick. Their internodes are also short with lots of branching space in between.

Cannabis has been in existence for a very long time. Our ancestors used it, mastered it, and applied it effectively in curbing and curing many diseases, defects, and deficiencies. This chapter is geared toward showing you the light as regards cannabis and helping you see the plant from a whole new point of view.

Legalizing the growing of cannabis would open a new path that will lead to development, change, and prosperity unlike the lies, propaganda, and stereotypes churned up to defame the plant.

Chapter Two

Seeds: What to Know

Like all plants, the cannabis plant begins with the seed. Presently, the cannabis plant seeds range to over 400 varieties and each seed is quite potent. However, some supersede others in terms of potency, growth, and the ability to thrive under harsh weather condition.

As a grower, you can either plant the seeds individually or a crossbreed of more than one species. But the question remains - how can you differentiate which seeds are the best? It's pretty simple. Along the course of this chapter, we will explain what you should look out for while selecting seeds.

Getting a seed with high quality is paramount. Having a conducive environment with adequate water, temperature, and sunlight contributes to a better harvest. But, getting good quality seeds is of great importance. You can either go for plain/pure seeds or the much-preferred crossbreed seeds.

The plain seeds are generally available in the market while the crossbreeds are limited because of the planting experiments they are used for by breeders. The classification of the seeds are as follows:

1. **Pure Sativa:** These are plain and pure species without no mixtures whatsoever.
2. **Sativa:** This species comprises of generally the mixture of Sativa and Indica.
3. **Pure Indica:** These are plain and pure species without no mixtures whatsoever.
4. **Indica:** This species comprises of generally the mixture of Indica and Sativa

5. **Indica/Sativa:** This is the equal combination of both Sativa and Indica species.
6. **Ruderalis:** These are species that are plain and pure.

Therefore, when you are in the market to purchase cannabis seeds, be sure to check the classification, features, and ratios of the crossbreeds so as to get the best quality. There are occasions where the combination of the crossbreeds might not be equal, thus making it's plants lean towards one side of the ratio more than the other.

Having an in-depth knowledge of this would give you an idea of what to expect as your crossbreed seeds start germinating. Notwithstanding, the grower can also exert an influence in keeping the growth, length, and look of the cannabis plant exactly how he or she wants it to be. The style, time, and technique we employ in harvesting our cannabis plant also play a part in influencing the high type.

Harvesting it later after it is ripe induces a couch-lock effect. On the other hand, harvesting the cannabis plant beforehand only leads to a cerebral high.

Choosing the Seeds

With over 400 varieties of seeds to choose from, getting the right seeds for your garden shouldn't be hard to pull off. However, knowing which cannabis seed is perfect for indoor and outdoor is paramount. In this case, we would recommend you stick to the breeder's instructions or advice.

To this effect, indoor cannabis seeds should be planted strictly indoors while the outdoor cannabis seeds should also be planted outside for much better results. Nevertheless, you might want to be a little more adventurous - and there is room for that. You can decide to interchange the seeds and substitute the indoor seeds for outdoor seeds and vice versa. The end result might even be much more productive in the long run.

After a careful selection of both indoor and outdoor seeds, the next best thing to do is to check the flowering times. Flowering times in the cannabis plant differ from one species to another. Naturally, flowering in the cannabis plant comes after the germination and vegetative growth of the plant.

Thus, endeavor to check the seed bank's instructions and manuals so as to stay abreast of all the necessary information about the seed you will eventually choose. For example, the seed bank might state that the period of 7-9 weeks after germination is the flowering time of the cannabis plant. This would give you an idea of when to start preparing for harvest.

The appearance and feel of the seeds are also paramount when choosing a seed. Seeds that look fresh, and healthy are always of good quality. However, with some seeds it is hard to tell from their appearance alone. Seeds of great quality exhibit coloring that is far darker on the outside.

Sometimes, they come in grey colors with a touch of tiger stripes. They are often firm and can stand the test of time. They are also hard as weak seeds end up cracking when pressure is applied to it.

Types of Cannabis Seeds

1. **Feminized Seeds:** This type of Cannabis seed is the most populous you would find around. According to research, 95% of the germinated feminized seeds end up yielding feminine cannabis plants. To this effect, there are lots of benefits associated with the feminized seeds. They produce buds, which makes pollination unnecessary. With more than enough space, adequate sunlight, and good temperature, the feminized seeds are bound to grow well and offer a bountiful harvest. The three best examples of the feminized seeds are Amnesia Haze, Afghan Skunk, and Yellow Lemon Haze.

2. **Autoflowering Seeds:** As a unique and improved cannabis seed, autoflowering seeds have the strength and ability to thrive even in severe weather conditions. The autoflowering seeds grow within the period of 10 weeks before the plant automatically blossoms with flowers, which indicates the beginning of its peak. They are a mixture of both male and female. Autoflowering plants are quite shorter than those from feminized seeds. The three best examples of autoflowering seeds are AK 47 AUTO, Amnesia Haze AUTO, and the Low Rider AUTO.
3. **Regular Cannabis Seeds:** These seeds germinate into male and female cannabis plants. Often times, they are equal in numbers, especially during harvest. Regular cannabis seeds were the first of the cannabis seeds; that was long before the Feminized and Autoflowering seeds came into being. As a traditional type of seed, they require more care and attention poured into them for a proper turn out.

Don't get it twisted, all cannabis plants need special care and attention poured toward them for a proper and perfect bloom. However, some seeds just demand it more than the rest. With over 400 seeds at your fingertips, choose wisely. Endeavor to weigh all possible outcomes and options before choosing a seed for your dope garden.

Don't be afraid to try a new mix of crossbreeds. It all depends on the kind of weed you want to grow. You can even try out personal seeds sourced from a friend's cannabis garden. The outcome might be mind-blowing. But the most important thing still remains, a good quality seed begets a strong and healthy plant. Now, let's move to the next chapter, shall we?

Chapter 3

Cultivating the Marijuana Plant

By now, I am sure you must be familiar with what cannabis entails and the kinds of seed you would like to use for your garden. As a grower who wants to start cultivating the marijuana plant, either as a beginner or a professional, you have to consider some stumbling factors as well as ask yourself important questions beyond venturing into it.

Cultivating marijuana is not child's play, especially with a large number of campaigns and crusades programmed against it. The government, to a large extent, has been able to restrict the growth and distribution of cannabis by arresting and prosecuting growers and distributors. The system seems to kick against it. Thus, cultivating a garden of weed takes courage, tact, and carefulness.

Hitherto, there is no actual offense to being caught with seeds, leaves, or any other parts of the cannabis plant. However, cultivating the plant is an unlawful act which is punishable by the law. What makes the sale of the plant punishable is the "trafficking offense" tag labeled on it - and it comes with no less than 10 years imprisonment. Cultivating cannabis is a serious offense.

So, before leaping into this business, we would suggest you take a good look as well. Ensure no stones are left unturned. Ask yourself these compelling questions:

1. Are you going to be around always to look after your garden of weed? If not, do you have someone trustworthy for that job?
2. How good is the security of the site?
3. Is it spacious enough to hide the smell?
4. How secure is your grow room?

These and more are important questions you should ask yourself before starting cultivation. Growing marijuana can be a lot easier than you might imagine. If all your answers to these questions are yes, then we would say you are one step away from being an expert grower. Be that as it may, a lot of people have been growing cannabis peacefully both indoors and outdoors without attracting a third party. Now, the question you should ask yourself is – are you ready to be a part of those smart men?

One rule all growers of the cannabis plant abide by is "Never Tell Anyone That You Are Growing Cannabis." This is the most sacred rule of them all; the same rule that has guided the privileged "Smart Men" to continuous, stress-free cultivation of the cannabis plant all through harvest periods. Additionally, in this line of business, you cannot afford to have loose lips around you. Like the saying goes, loose lips sink ships.

Let's take Simon, for example:

Simon is a fun loving person with a deep passion for cannabis. Once in a while, he smoked it along with his friends during parties or home picnics. He wanted to start growing his own weed instead of getting it from the street, but he had no idea how to even begin. Then Boom! He came across the "Growing Marijuana Handbook." He had since become an expert grower in his own home.

Don't get it twisted, Simon shares his weed - but with a cover-story. "Hey, Guys! Look what I just bought downtown?" "A friend just gave me greens as a gift for a job well done. Let's share." These stories and more have saved Simon's life and garden a million times. My point is: **Be. Like. Simon.** Some of your friends might not be good at keeping secrets. Keeping your garden away from the public is what will help you become a better and expert grower.

Cultivating cannabis is a serious business that must be handled with passion and enthusiasm. You must be willing to treat your garden like it's a part of you. That is the only way you will give it the necessary care and attention it needs to thrive. Regardless of the kind of species you are growing, the kind of special seeds you are sowing, and the kind of improved cannabis plant you want to cultivate, if you don't take good care of the plant, it will eventually die off.

Cannabis Life Cycle

Cannabis is unique. It is special. Unlike most plants that feature a 6 stage life cycle, the cannabis plant has just 3 main stages. Want to know them? Here they are:

1. **Germination:** Germination is the first stage toward the cultivation of cannabis either in a large, medium, or small scale production. You can't just assume getting the seeds is enough. No wind, birds, or animals are going to fly or walk through your garden and accidentally start sowing your seeds for you. And trust me, sowing can be quite fun.

    Putting a seed or two in the soil is always the first thing to do. With much care and proper gardening, the seeds crack open, bringing forth a newborn seedling on the surface of the soil. Its root begins to spread like tentacles, catching a firm hold of the soil beneath. This gives balance to the seedling whilst it shoots up.

    When sunlight touches this young seedling, it paves the way for fresh leaves, thereby pushing the seed's shell far away from itself. This is the beginning of life for a cannabis plant. Whether it will grow into full maturity from here now solely depends on the type of care, attention, and techniques you employ afterwards. Naturally, this germination process takes place in the span of three weeks.

    Now guess what happens afterwards? Tons of fresh leaves end up sprouting with quite a number of marijuana characteristics. The leaves start forming finger shapes, the edges start getting pointy, and the stems start getting thicker. This is known as the seedling stage. I would advise you use a stick or a wooden stake to hold the seedlings in an upright position. Replace the sticks if you have to, because this phase lasts three good weeks, setting the pace for the plant to sail through to the second stage – Vegetative Growth.

2. **Vegetative Growth:** The vegetative growth of any cannabis plant irrespective of the species is usually the same. Little by little, the seedling begins to gain energy from the adequate sunlight, temperature, and water it comes in contact with. With time, the leaves begin to get bigger and the stem becomes thicker. Telling the sex of the plant would be quite glaring at this stage of the cycle.

   Most times, the vegetative growth stage also marks the beginning of the pre-flowering period. Ensure you are well-prepared as your marijuana plant hits the road to maturity.

   The pre-flowering phase paves way for the flowering stage. One distinctive feature during this phase is the slowed-down growth of the plant. This stage triggers a new development which focuses on filling out, instead of growing tall. Also, calyx will appear at strategic points. With this change, the stems are bound to grow thicker, leaves wider and fuller, and the branches sprout out even more.

   Clip off unwanted parts. By now, your plant must be able to stand upright without sticks or stakes. No doubt, the vegetative growth stage is the pathway toward our next stage of the cycle – the Flowering Stage.

3. **Flowering Stage:** The flowering stage comes right before the harvest period. You might want to hold on tight to something as you behold the beauty and sight of the blooming flowers of the cannabis plant. The flowering stage showcases full-fledged, beautiful flowers that will surely blow your mind away.

   The sex of each plant becomes more glaringly obvious at this stage. The male plant will showcase a collection of grape-like balls while the female plant will feature whitish pistils around its leaves. However, flowers in the cannabis plant take time to fully develop.

To this effect, something amazing will happen as the plant keeps pushing towards maturity. The male plants' pollen sacks will reach a point where they will burst, thereby dispersing the pollen to the female flowers. In turn, the female plant will reciprocate by making seeds out of that pollen. The seeds stay in the female buds for weeks before they attain maturity.

You might be surprised as a beginner when your plants start dripping liquid. No need to worry, as your plant is approaching maturity. Those liquids are called resins. Resins are formed as a result of the large size of the buds which would make your plant very sticky when you touch them.

The cannabis plant growth system can be likened to the normal system of an animal in heat. Wondering if that makes sense? Then here is my comparison. Animals in heat showcase their need for an opposite-sex partner, in order to copulate amongst themselves. Failure to meet with one leads to not-so-good reactions like aggressiveness, stress, and abnormal behaviors from the animal.

This is the same way the female cannabis plant reacts to the absence of pollen from a male plant. Instead of being aggressive, stressed, and abnormal, it produces resins from the big, swollen buds, which are trying hard to attract the male pollen. This signals the beginning of harvest and full maturity of the cannabis plant.

How do you know they are mature? It's pretty simple! The white female pistils change colors, thereby opening the seedpods automatically. During this phase, it is advisable that you stay alert, so you can collect the seeds yourself before they end up being dispersed on the floor.

Cultivating cannabis means tending to the plants, watering them at the necessary time, weeding out an unwanted plant that may hinder its growth, and so much more. It's a total devotion that must be adhered to.

There you have it! Cultivating cannabis at your fingertips! Like you were promised at the beginning of this Handbook, by the time you finish reading through it, you will be more enlightened about the cannabis plant than when you first picked it up. Don't forget to space your cannabis plants. That way, they won't be competing for adequate air, water, and sunlight, which might, in turn, lead to unbalanced growth in your weed.

Cannabis cultivation varies in techniques. The indoor growing style is quite different from outdoor. This and more are what we will explain in the next chapters. Discover how to properly start up an indoor dope garden. You wouldn't want to miss it. And remember, always be like Simon - courageous, tactful, and very careful.

Chapter Four

Indoor Growing: Setting Up the Indoor Garden

Having learned about the basic concepts, history, seedlings, and steps of cultivating cannabis in the previous chapters, the next thing to focus on is one of the core production stages, which is transferring your seedlings to the permanent growing area. This area or environment will be the home for your seedlings for the next 3 to 6 months.

In Chapter Three, we asked vital questions as regards growing cannabis. If answered correctly, this would ease the stumbling block between you and the success of your dope garden. Choosing a conducive, quiet, nice place for your garden can be quite hard to pull off. However, there are necessary factors you would have to consider before choosing a great site.

Before then, a decision needs to be made! Are you planning on cultivating your cannabis plants indoor, or growing them outdoor? Ask yourself this question. If your answer is the former, then we would suggest you read this chapter with rapt attention. Indoor cannabis gardens started decades ago with the heavy price placed on the necks of cannabis cultivators. Growers of cannabis had to crawl back into their shells, if their lives and businesses meant anything to them.

Crawling back means only one thing. Re-strategizing and re-planning. Thus, indoor cannabis cultivation was born. Law enforcement agencies could no longer smell their presence even from afar. Indoor growing became the only way to keep growers' heads up sitting on their necks.

Cannabis growers no longer had any other option other than to focus, experiment, and perfect indoor gardening. Indoor gardening, to a large extent, helped lots of growers carry on with their activities with little or no worries from the authorities. Indoor growing gives them the perfect cover they crave.

Additionally, indoor growing simply means what the term implies; growing a plant or crop indoors. This can be either at home, at a business area, or in an enclosed environment. This method of farming involves the use of hydroponics and artificial or natural lightning to provide the essential nutrients needed by the plant. Cannabis cultivation thrives under this indoor method of growing cannabis.

Setting Up the Indoor Garden

Now, setting up the permanent area or environment to grow your product is essential. No doubt, there are lots of factors to consider, issues to tackle, and problems to address. Amongst these important factors lies privacy and security. In order words, privacy and security, to a large extent, play a very big part in choosing the site.

How secure is the environment? Is it a secluded area? How many people know about the site? Can it withhold the smell when it blossoms? How secure is it when people visit you? These and more are questions you should ask yourself before setting up an indoor dope garden of your own.

Like we said earlier, the factors to consider are quite numerous. In order to set up a state-of-the-art indoor garden, no matter how low your budget is, tackling these factors that would end up causing obstacles to your precious plants is of great importance. This will help prevent failure in your quest to grow good and healthy cannabis indoors.

There are many factors, like your financial capability, the right lighting to use in order to avoid overcooking the cannabis, the right place to show your cannabis seeds with enough privacy and security, etc. The factors are numerous, thus discussing each and every one of them would be impossible. Therefore, we would pick just a few. Follow us, as we will be discussing more details on these factors below:

1. Lighting

Lighting is the most important factor to consider when growing cannabis indoors, and there are two major ways to generate light for your products. These are natural lighting and artificial lighting. These two ways, even when perfect for an indoor garden, come with their strengths and flaws. For instance, if you are looking to produce bigger flowers (more buds), then artificial indoor lighting is the best. But if you are looking to spend less, natural window lighting is perfect for you.

Still not following? Then let's break it down for you. Artificial lighting yields more product but it's very expensive to run. While its counterpart, natural lighting, is relatively cheap, it offers fewer yields. Thus, if you are trying to run the artificial lighting method of growing cannabis indoors, then it is very important for you to consider your electricity bills and the kind of light to use.

Hence, you have to think of how much electrical power will be consumed. How much heat will be generated from the lamps? And how much light will be emitted from the lamps? Modern-day cannabis growers tend to use three major artificial light sources that are very effective. These are:

1. High-Intensity Discharge Lamps (HID).
2. Compact Fluorescent Light (CFL)
3. Light Emitting Diode (LED).

These lights can be used separately or in combination, but you have to be careful how you combine the lights so as not to overcook your cannabis.

On the other hand, if you are looking to run much cheaper cultivation of cannabis, which equates to using the natural window lighting method of growing, then taking into consideration the privacy and security of your cultivation are of great importance. Light to the cannabis plant is one important nutrient it needs to thrive. And the natural lighting method of indoor cultivating can only be accessed via sunlight, or window light.

To this effect, it is important to place the cannabis plant in a strategic position near a window where light can easily reach and penetrate the plant. To get the strategic position you need, you have to note that the sun rises in the east and sets in the west, just as it travels north or south depending on your location. As a result of this, you have to study the sun's position in your location in order to achieve a desirable result.

Before you do that, though, you have to make sure your cultivation is safe by considering whether people can look up and see cannabis by the window or if people who wash your windows can easily spot the cannabis through them Also, note that the sativa plants usually grow very big and can be easily spotted. So it is important you take these factors into consideration before delving into the natural window light method of growing cannabis.

2. Growing Space

Another factor to consider when running an indoor method of growing cannabis is your growing space. This could be a whole room or part of a room dedicated to the cultivation. Places like the basement, attic, spare bathroom, hot press, and closet are good locations to start cannabis cultivation. You can even decide to build your own cabinet specifically for the cultivation. Also, note that some of these locations need artificial light to enhance growth.

3. Time

Timing is an essential factor when it comes to indoor cannabis growing, especially when you are running a natural window lighting cultivation. This helps you determine and plan toward the season when the sun is at its peak. For instance, if the sun is at its peak in the month of July, then it is advisable you start producing seedlings as early as late March, April, and May.

Therefore, it is essential and very important for a cannabis grower, in general, to be able to predict and guess when they will get the best weather. This factor helps the grower foresee the harvest time for their cannabis cultivation, especially when they coordinate this with the flower times as instructed by the breeder.

4. Soil

The soil is another factor to consider when setting up indoor cannabis growing. For the fact that you can't cultivate cannabis outside on a plantation or farm because of security and privacy, and buying the expensive Hydroponic Substrates is also not an option, finding the right soil for cultivation is the next best thing to do when setting up your indoor growing. As a matter of fact, it doesn't cost a cent.

Nevertheless, you just can't jump at any soil you see in your environment. Fertile or not, not all soils work hand-in-hand with cannabis. Like we said in our previous chapters, cannabis is special and it is unique. Thus, here are the three things to consider when choosing a good soil in order to grow healthy plants:

1. pH
2. Structure
3. Nutrients

Also note that the use of soil in growing gives your cannabis a certain taste and aroma.

Setting up an indoor cannabis garden is very easy if you are willing to take the chance. With determination, planning, and dedication, your weed garden will be up and running in no time. Just ensure all boxes are checked, no stones are left unturned, and all measures are taken for a perfect cannabis cultivation indoors. Join us in the next chapters where more of these factors will be followed up and discussed in clearer details.

Chapter 5

Indoor Growing: Planting and Planting Techniques

Indoor growing of cannabis is very easy to practice when you have the right set-up and I am certain you already have an idea of what your indoor garden should be like from the previous chapter. In this chapter, we will be focused on planting properly and the planting techniques you can use to achieve a masterpiece indoor garden.

There are so many ways to grow cannabis indoors, and selecting which to practice is totally dependent on what you can afford and what is best for you. Moving forward in this chapter, we will be focusing on the two major methods or ways of growing cannabis indoors. These ways are Soil Growing and Hydroponics.

Soil Growing

Soil growing is one of the easiest and least expensive methods of growing marijuana indoors. With the right soil growing set-up, you are sure to cultivate and harvest a delicious and amazing crop of cannabis. With an effective light system already in place, the next thing to focus on is the right soil combination for your soil growing set-up.

Choosing the Right Soil for Indoor Soil Growing

Choosing the right soil for your indoor growing is very important as there are many types and varieties of soil out there to choose from. But this shouldn't be a problem, because as we proceed, you will find out that different ingredients can be added to different soils to achieve a healthy and improved cannabis plant growth.

When growing cannabis, the soil you use goes through different phases and periods. In the first period or phase, the soil contains water and other nutrients which the cannabis' long, winding roots absorb in order to grow. During the other period or phase, the soil becomes dry, and this could be as a result of the absorbing of most of the soil water by the cannabis plant root or as a result of you not watering your plant.

Do not panic, because in these dry periods, air finds its way into the soil and allows the roots of your cannabis plant to breathe. As we mentioned in the previous chapter, there are three major factors to look out for when choosing a soil for cannabis indoor growing. These are the pH, soil structure, and nutrients. Here's why these factors are important and essential for your indoor growing soil.

Factors to Look Out For

1. **PH:** Soil pH helps you determine the acidity and the alkalinity of a soil. A soil pH is measured on a scale from 0 to 14, where 0 is very acidic, 7 is neutral, and 14 is very alkaline. The pH of soil is usually written on the bag. Now, for growing a cannabis plant, you need a soil that is neutral because if it is above or below 7, it can cause problems to the growth of your cannabis plant.

   Therefore, go to any gardening shop and get a small pH meter so you can measure the overall pH of the soil you are going to be using for your indoor soil growing.

1. **Nutrients:** There are three basic nutrients found in soil. These are: Nitrogen, Phosphorus, and Potassium, popularly known as NPK. The percentage ratio for NPK is usually 20:20:20. In other words, 20% Nitrogen, 20% Phosphorus and 20% Potassium. The remaining 40% represents other particles that make up the soil, including liquid, which is usually water.

   This information can be found on the bag of the soil you are buying. It is also important to note that these ratios can change because of different types of soil brand and nutrients, so it is essential to know the right percentages of the different nutrients your cannabis plants need. Based on research, we have come to understand that cannabis plants require a higher level of Nitrogen, and normal levels of Phosphorus and Potassium during vegetative growth.

Therefore, when picking out a soil, you need to pick a bag that contains all three nutrients. This is also applicable when picking out chemical fertilizers. Look for a mixture with the highest first number or a first number that is equal to the other numbers. In other words, the ratio of 12:12:12 or 20:20:20 is fine and the best you can find out there. 12:6:6 and 18:4:5 are also fine and work great for your indoor grow.

A ratio like 8:20:20, on the other hand, is not suitable for cannabis plant growth. This is because soil with a higher Phosphorus level is best for flowering plants, not cannabis.

2. **Structure:** The best soil structure to look out for your indoor soil growing is a soil that is not too moist. Hence, pick out a soil that easily dries out and doesn't hold water like a muddy soil, as this can sometimes cause damage to your plant because of the lack of air. The roots of the cannabis need air to breathe.

Also, pick out a soil that is not too dry and will not need to be watered very often. In other words, you need to find a soil that balances the two (wetness and dryness), a soil that is loose but feels fine and kind of heavy when touched.

Choosing a soil is totally up to you so you have to be careful not to pick out a soil that is too soft and weightless or too hard and bulky. Medium soft and heavy is the kind of structure your cannabis plant seeks.

Soil Types Suitable for Growing

At this point, it is important for you to note that you cannot just bring in natural outdoor soil for the cultivation of cannabis because the soil may end up not being sterile and may contain pests and bugs that will affect your plant. It is advisable for an indoor grower to always purchase soil from a garden shop. In this part of the chapter, we are going to be discussing some of the soil types and why they are suitable or not very suitable for planting cannabis.

There are many different types of soil medium out there that marijuana can grow on, but some of these soils may not be very suitable for marijuana cultivation. Below is a list of the different types of soil and their effect on marijuana.

1. **Sand and Silt soils:** Sandy soils are usually pure sands or a mixture of sand and soil. As a result of this, sandy soils are very dry and easily drain out water and other minerals too quickly from the soil. This soil is therefore not suitable for our needs in growing cannabis.

   On the other hand, silt soils are like sandy soil but darker and more clay-like. They have the ability to hold other nutrients but cannot hold water like sandy soils. In essence, sandy soils and silt soils are not suitable in our quest to achieve a successful indoor marijuana garden.

2. **Clay Soil:** Clay is a stiff fine-grained type of soil that contains hydrated aluminosilicate which makes the soil flexible when water is added. This type of soil is certainly not very suitable for the growth of marijuana and is rarely used for its cultivation. It is only acceptable to use this when it is mixed with other types of soil.

3. **Humus:** Humus soil constitutes organic matters which are created by the decomposition of plants. This type of soil is also sometimes known as compost, which is the final mixture of manure from organic matters. Loamy soil and other mediums are added to complete the mixture of this soil. Humus soil can be purchased from any local gardening store.

4. **Loamy Soil:** Loamy soil is a mixture of all the above three types of soil; sand, silt, clay, and humus. This type of soil is very fertile and highly recommended for your indoor growing. The information containing the mixture of the soil is always found on the bag and it is very likely you will find this type of soil in a local shop.

Choosing the Right Pot

Having chosen the perfect soil for cannabis cultivation from a local gardening store, the next thing to look out for is the right pot. Planting pots come in different sizes, shapes, colors, and abilities to add certain benefits to your plants. When choosing the right pot to grow your marijuana in, you need to look out for a much larger pot of about 1.5 to 3-gallons because cannabis tends to develop long roots.

Another thing to look out for when picking the right cannabis planting pot, especially if you are a newbie in indoor cannabis growing, is perforated pots. These are pots that come with holes or perforations at the bottom. These perforated pots sit on a dish which helps retain drained water from the soil after watering. You are expected to change the dish after a while.

The reason for these perforated pots is to control and drain excess water from the soil, especially when the plant is overwatered. Overwatering is very bad for your cannabis plants as it can kill your plant, waste nutrients, and cost you time. In other words, you have to be extra careful not to overwater your plants.

After getting your pot, it is expected for you to clean it properly in order to rid the pot of all unwanted chemicals, particles, or dust it has gathered over time. Also, it is best to use one pot per plant, so if there's a problem with the plant, nutrient, soil, and watering, then it will affect only one plant.

Indoor Soil Growing

There are a lot of options to consider when you want to cultivate cannabis indoors. Below, we are going to list and explain 4 most common indoor setups. These are:

1. **Bench growing Technique**
   Bench growing has to do with growing cannabis on a bench or table which is made of aluminum or wood. The pots where the plants are cultivated will be placed on the bench or table for growing. The

bench growing method is mostly used by commercial marijuana producers.

2. **SOG (Sea of Green) Growing**
SOG growing or Sea Of Green growing is a process of growing cannabis where, at a young age of cultivation, flowering is forced. This method of growing enables you to harvest earlier than the normal harvest time and allows you to maximize space without sacrificing your yields. This method also enables fitting more flowering plants into a small space.

SOG Growing Technique Steps

1. Germinate seeds or create clones.
2. With 18 to 24-hour lights, grow your plants to reach 10 or 12 inches in height.
3. Space your plants and increase lighting regimen to 12/12 in order to force flowering.
4. When you finally achieve a dense canopy, trip branches from bottom to use them for the clone.
5. Harvest!
6. Grow again.

3. **ScrOG (Screen of Green) Growing Technique**
The ScrOG or Screen of Green growing technique has to do with growing marijuana indoors with the aid of screens to maximize light exposure and horizontal growth. Chicken wire or nylon netting can be used to influence the movement and growth of marijuana on the screen. It is expected that your screen has holes with a diameter of 5cm. This method is mostly used when your grow room has limited space for growing.

ScrOG Growing Technique Steps
1. Place your screen 20 to 25 inches from your plants.
2. Cut the top cola off of each plant.

3. When the top of each plant reaches the screen, prune bottom branches.
4. After two days of pruning the bottom, force plant into flowering.
5. Use chicken wire or nylon netting to attach all branches to the screen horizontally.
6. Start growing!

4. **Cabinet Growing Technique**

The cabinet growing technique is another easy and efficient way of growing cannabis indoors. If you have limited space or are just trying to experiment as a newbie or beginner, you can start by practicing cabinet growing. This entails growing your cannabis plant in a small space like a closet, cabinet, growth tent, or grow box.

Cabinet Growing Technique Steps
1. Clean and clear a cabinet or closet space properly ridding the space of molds, dust, and dirt.
2. Paint inside the cabinet or closet, with preferably white paint, or use a strong white plastic to cover inside the cabinet.
3. Set up the lights.
4. Put your potted plant in place.
5. Set up fans and carbon filters.
6. Start growing!

Hydroponics

Hydroponics is another efficient and effective way of growing marijuana indoors if you are not looking to use the soil growing method. Hydroponics is the growing of plants in gravel, sand, or liquid, without the presence of soil. Nutrients are added in the process of hydroponics to increase growth.

In other words, when growing marijuana with this method, soil is not going to be used and the medium for plant growth is definitely going to be sterile and inert. Plants grow faster because this method provides your plant with nutrients, air, and water. The hydroponic method of indoor growing is really great to practice and is ideal for areas that are frequently in drought.

From the previous discussions, we have come to understand that marijuana plants require certain nutrients to promote its growth and quality. Without these nutrients, it is very likely that your plants will die. Under the soil growing method, we came to the conclusion that the soil provides the marijuana plants with nutrients like NPK for better growth.

It is also important to know how to add nutrients to your plants when practicing the hydroponic method of growing. This can be done by creating a balanced nutrition solution containing Nitrogen, Phosphorus, Potassium, Calcium, sulfur, and magnesium mixed with water. This method also makes it easy to adjust nutrient levels for the different stages of growth.

Just like the soil growing methods, there are different hydroponic techniques you could use for your indoor growing. These are:

Ebb and Flow Method

This is the simplest and most common method of hydroponic indoor growing. It is perfect for beginners. This method involves the use of 3 reservoirs, one holding the water and nutrients needed for your growing, the second holding clean tap water that can be reserved for 2 to 3 days, and the third as a spare in case either of the other two breaks when in use.

The reservoir you will be using in this method should have a lid and be insulated so you can control the temperature of your nutrient solution

Aeroponics

Aeroponic is a method of hydroponic growing that doesn't need any growing medium. The roots in this method hang in mid-air, which is kept at 100% humidity with a sprinkler that sprinkles most of the solution, enabling the plant to absorb the nutrient while oxygenated. This method can lead to faster growth (up to 10 times faster than soil) for your plants.

Continuous Flow Method

Continuous flow refers to the continuous flow of nutrients to the roots of a plant. This method involves the use of pipes with holes at the top where the potted plants are placed and constantly supplied nutrients to the root inside the tube.

After reading this chapter, it is certain that one can easily set up and start an indoor growing technique or method without much trouble. Hence, starting an indoor planting or growing is totally dependent on the technique you choose - this being either the soil growing method or the hydroponic method.

Stay with us as we delve deeper in the next chapter on the right nutrients and environment suitable for your indoor growing.

Chapter 6

Indoor growing. Nutrients and Environmental Factors

With all our discussions in the previous chapters, I am certain you can now create an ideal environment and provide the right nutrients for your cannabis plant. The growing success and failure of your plant is totally dependent on you as a gardener, because you have the ability to fully control the environment and the number of nutrients provided to the plant.

We will be discussing in this chapter two major factors that are essential and contribute greatly to the development and growth of your marijuana.

Nutrient Factors

We have come to understand that nutrients are one of the most important factors needed for your cannabis cultivation to strive, grow, and develop into a healthy plant. These nutrients are classified into three categories.

The first category is the basic nutrients, otherwise known as the primary nutrients. These nutrients are Nitrogen (N) Phosphorus (P) and Potassium (K), or NPK as popularly called. We have talked about these three basic nutrients in previous chapters and come to the conclusion that they are instrumental to the survival or death of your cannabis plant.

The next category of nutrients found in the soil is secondary nutrients, or supplements as they are sometimes called. Secondary nutrients are also found in the soil, but in some cases, not all of these three supplements are present. These supplements are Calcium (Ca), Magnesium (Mg) and Sulphur (S). They also contribute immensely to the growth of your cannabis plant such that if any of the supplements are missing in the soil composition, it could lead to a deficiency in your plant's nutrition.

The last category of nutrients found in the soil is micronutrients. They are seven in number, namely: Zinc, Manganese, Iron, Boron, Copper, Molybdenum and Chlorine. These micronutrients don't have much effect on the good health and growth of your marijuana plant and problems of micronutrients hardly come up except there is a chemical reaction between the nutrients in the soil. This situation is referred to as a Lockout. Lockout takes place when a large percentage of one particular nutrient is being applied to your plant. This act can cause an imbalance between the nutrients and lock out other nutrients from being used. When this happens, your soil will need to be flushed.

In a situation where your soil faces nutrient deficiency, probably as a result of your plant absorbing most of the nutrients, the best solution to tackle this is feeding your plant.

Feeding

Feeding in this case has to do with refilling the soil with missing nutrients that have been absorbed by the marijuana plant. Feeding should only occur when your plant actually needs to be fed. In most cases, feeding your marijuana plant is done after 14 days at less than 50% of what is written on the nutrient label, in order to avoid plant burns. This is because marijuana plants easily experience burns, even with a small amount of nutrient feed. For instance, if you read on the label of the nutrient feed that you should use one full cap to 3 large gallons of water, you are advised to therefore use one full cap to 6 gallons of water.

Throughout the life cycle of your marijuana plant, you are expected to use only three solutions to feed your plants. These solutions are

1. A nutrient feed solution that has equal NPK or a solution that has more N over P and K. This solution is used in the vegetation growth stage of your cannabis plant.
2. A nutrient feed solution that has more P over N and K. It is used in the flowering stage of your plant.
3. A third nutrient solution is a solution that contains the secondary nutrients.

Note that secondary nutrients or supplements should be added every three weeks.

Air

Fresh air is essential in the healthy growth of a marijuana plant, especially when the plant is in the vegetative and flowering stages. Do not forget that there will come a time when your cannabis plant will experience dry periods in order for its roots to breathe. This just shows you how important fresh air is to your marijuana. Be sure to always open your windows for fresh air as it is the best option for your plants.

In winter, though, open your window for just 15 to 20 minutes to avoid freezing and stunted growth; and if you are growing in an enclosed area, be sure to create a good fanning system where your plants can get refreshed air. This can be done by using two fans, one to extract old air and the other to provide refreshed air.

Humidity

Humidity is simply the percentage of water available in the air. 40 to 80% relative humidity is what is best for your marijuana plant. Do not use equipment like a dehumidifier to control the humidity level of your grow room except if you are running a large operation. Be sure to always use fresh air to control and maintain the humidity level.

Temperature

The temperature of your grow room also has an influence on the healthy growth of your cannabis, hence the need for natural sunlight or a heating unit. Household temperature is therefore acceptable to grow cannabis in.

It is important to know the temperature that is suitable for your plants. Therefore, you have to learn to control the temperature level of your grow room. The best way to measure your grow room temperature is the use of the human body. If your grow room is too cold for you, then it's certainly too cold for your plants and therefore not suitable to grow your cannabis in. So it is advisable to utilize a thermometer in determining the temperature of the room just like the human body. Warm room temperature of about 75 degrees Fahrenheit should be the ideal target for your grow room. Therefore, when the room is too hot, open the windows or turn on the fan. When it's too cold, turn on the heater.

Fan

There is so much emphasis placed on the use of fan especially when you are practicing indoor growing. The slight, fresh breeze produced by the fan is important because it adds strength to the branches and stems of your cannabis plant. It also gives an outdoor feel to your indoor grow room.

Carbon Dioxide (CO2)

During photosynthesis, plants take in CO_2 and release O_2, but there is no balance between the intake and release of the two elements, because the CO_2 is essential to the survival of your plant and helps the plant grow bigger buds. It is advisable that you as a grower use a carbon dioxide generator to increase your plant intake of CO_2.

Environmental Factors

Environmental factors have to do with controlling and maintaining the soil and pH level of your plants. These factors also influence the success and growth of your marijuana plants.

Soil Control

Soil control has to do with the removal of waste materials added to the soil by the cannabis plant and the addition of nutrients that have been absorbed by the cannabis plant. This process can lead to fluctuation thereby decreasing and increasing the pH level of the soil. Now it is up to you as a grower to maintain the pH level to seven or neutral in order to avoid an acidic or alkaline soil. This process is better known as soil control.

pH Control

pH control has to do with checking and maintaining the pH level of your soil to always be neutral. We have stated in previous chapters that when the pH of soil goes a bit lower or higher than seven (neutral), your soil pH needs adjusting. Therefore, it is advisable to check the pH level of your soil once a week or two days after feeding your cannabis plants.

One of the best ways to adjust your soil pH level when it has moved from being neutral (7) to acidic or alkaline is by using certain ingredients to bring the soil pH back to its normal level or engaging in a process called soil flush.

Bringing the pH Level to Neutral

This method consists of two processes: the bringing of the pH level from acidic to neutral and from alkaline to neutral. This method involves a lot of trials, mistakes, and errors, but this shouldn't stop you from experimenting. Some of the advanced cannabis growers today got where they are by a lot of experimenting and making mistakes. This method also helps you know and understand the number of ingredients and nutrients to add to your plants over time.

Also, note that your soil outcome after using this method will become impeccable.

1. **Acidic to Neutral**

    Bringing a soil's pH level from acidic to neutral has to do with the use of lime. Apply lime to the affected soil consistently, especially after watering the plant. As you apply constantly, the quantity of the lime you apply should reduce so the pH level reaches a neutral level where you eventually stop the application.

    Be sure to constantly use a pH meter to measure the pH level as you apply the lime.

2. **Alkaline to Neutral**

    To bring a soil pH from alkaline to neutral, you need to make the soil more acidic; and this can be done by adding certain ingredients to the soil. These are:

    1. Ground coffee
    2. Lemon peel
    3. Cottonseed meal

4. Acidic fertilizers

This method involves moderately adding any of the above ingredients until the soil eventually reaches a neutral level.

The products a lot of people, especially growers, are using today as a pH agent to adjust soil pH level is **pH up** and **pH down**. They are very effective and are available in different grow shops around.

Soil Flush

A soil flush is not highly recommended, but only ideal when your soil is experiencing serious fluctuations in pH level or when your plants experience serious burns caused by chemicals. It is highly advisable to mix raw nutrient feeds with water before using them on your plants, to avoid situations that will lead to soil flush.

Soil flush is usually a last resort and can be done in 3 different methods.

1. This method involves tilting your plant and pouring out the unwanted liquid.

2. This method is recommended for plants with perforated pots. In this case, you are expected to pour in plenty of water into the top of your pot and wait till the water flushes out from the holes at the bottom of the pot.

3. This method is highly recommended for plants without perforated pots. Either you create a perforated pot by drilling holes underneath it and follow the second method or you carry out a transplant immediately to new soil.

With knowledge and research, you do not need an expert to help you check the nutrient and environmental factors of your plants or even your grow room. Keep reading through the next chapter and be enlightened on how flowering and breeding works in a marijuana plant.

Chapter 7

Indoor Growing: Flowering and Breeding

If you have been consistently following all our discussions in the previous chapters, I am sure you have been able to set up a convenient and healthy environment for your marijuana plant. At this point, you should have a grow room with conducive fresh air, plants in pots under grow lights, balance in soil nutrients and pH, adjusted lights, flat and stretched out leaves, and finally an increase in the height of your plants.

If you have not been able to achieve some of this, you shouldn't worry, mistakes are inevitable and they only make you more knowledgeable in your field. So moving on, if you have noticed some of the above traits like flat and stretched out leaves and increase in plant size, be sure to note that the next thing to follow is the stoppage in the growth of your plant and the development of new growth on your plants, especially between the stem and branches.

Do not panic when you see this happening. This period in the life cycle of your marijuana plants is known as the end of vegetation and pre-flowering stage respectively. In order words, the end of vegetation is the growth stoppage of the marijuana plant and the pre-flowering stage is the period where the development of new growth on the marijuana plant takes place. This new growth will eventually produce more leaves, flowers, and branches.

Pre-Flowering

At this stage of pre-flowering, you should hope your plant produces more females than males and you will be able to tell the sex of your plants by using three major methods. These methods are not 100% accurate in the reveal of the sex of your plants but sometimes they serve as reliable indicators. These methods are:

1. **First Early Sexing Method**
   The first early sexing method is applicable when you are growing all your seeds at the same time. When these seeds start growing,

then you will be able to tell the sex of the plants by observing their heights. The taller ones are male while the much shorter plants are female.

Another thing to note in this method is that the male plants tend to start pre-flowering before the female plants. So when you notice a tall plant that started pre-flowering before the short plants, be sure to know it's a male plant and vice versa.

2. **Second early sexing method**
   The second early sexing method has to do with checking the calyx of the plant by using magnifying glasses. Now if you notice that the calyx is raised on a stem then it should be a male plant but if the calyx (new growth) is not raised on the stem then it is likely to be a female plant.

3. **Third early sexing method**
   This third early sexing method is known as force-flowering a cutting. It is the best method out there you can use to determine the sex of your plant in an early stage. This method involves cutting a piece of your plant and placing it in a cup of water or cloning medium like Rockwool for 24 hours.

   In this 24 hours, provide your plant with 12 hours of light and 12-hour darkness.

   The result of this method will be a flower displayed by the cutting. This flower will reveal the sex of the plant. If you end up using a cloning medium, then the result will be an exact clone that consists of similar genes and sex as the plant it was cut from, hence revealing the sex of the original plants it was cut from.

Like I said earlier, these methods are not seen as 100 % accurate and should not be totally relied on. These methods are advantageous as they help you understand differences between the male and female plants.

Pre-Flowering Period

The pre-flowering period of a cannabis plant is usually one or two weeks. In this period, the calyx region of the plant begins to take different shape and size based on its sex. It is during this period when the calyx is taking different shapes that you can fully determine your plant's sex.

Facts About Growing Cannabis

Making certain decisions sometimes can be difficult and without proper advice or research, you might end up making a decision that will affect you negatively. So, before you decide whether to flower your plants or continue vegetative growth, here are some interesting facts about growing cannabis that you should consider:

1. Some cannabis plants have a very long life cycle and can survive up to 10 years by just being under steady light at all times. Over this long period of time, the plant reaches its maximum height, stops producing branches, and eventually starts to turn to a bush. The rest of its life cycle will be focused on replacing old leaves by growing new ones.
2. Another interesting fact to note is that bud production is actually not equivalent to the height of your plants, rather it is equivalent to your strain's genes, the number of nodes your plant has, and your growing environment. It is also important to note that all plants' node areas will eventually turn to a budding area and every strain has the potential to produce buds.
3. Also, flowering a plant immediately after it has matured tends to increase the production of buds, especially if the plants are plenty, rather than the extended vegetative growth. In other words, the shorter option tends to produce more turnover in buds than the extensive option.

Flower Time

If your plants manage to reach the pre-flowering stage, it means your plants have matured enough to begin flowering. Now the most important question to ask is if you, as a grower, want to take a long route by continuing with vegetative growth or start flowering.

If you choose to take the long route, then you have to make provision for a bigger grow room because the cannabis plant tends to increase over time in height and width.

On the other hand, if you choose to flower your plants, then the best option is to utilize a 12/12 schedule for your cannabis plant.

The 12/12 Schedule

The 12/12 schedule promotes great outcomes and good quality buds for your cannabis plants.

This schedule has to do with providing your plant with the natural 12 hours of light and 12-hour darkness that a full 24 hour day provides. If you are using an indoor grow method, it is therefore important to switch on your lights for 12 hours and switch them off for another 12 hours.

This, therefore, gives your plant a natural stimulation to produce a flower. As you continue to keep up with this 12/12 schedule, your plants will continue to be plentiful with flowers, which is exactly what we seek as growers.

Also, note that for the success of this 12/12 schedule you are expected to commit fully to these routines. For the 12 hour darkness, be sure not to let any single light penetrate through the darkness because it can affect the plant's flowering process. In other words, your grow room should be totally and completely sealed from light. You can use a photographer's darkroom as an inspiration for creating your grow room.

12/12 Schedule Problems

If you start using the 12/12 schedule before your plants reach the pre-flowering stage, then you might face two major problems:

1. Sex problems caused by sex (the issue of hermaphrodite)
2. Abnormal bud growth.

Breeding

If you really want to start breeding your own cannabis strain, but you find scientific write-ups highly confusing, and graphs and punnett squares put you to sleep, then this is the handbook for you. We will break down how to breed your own special strains with more than enough traits to help them out in severe weather conditions.

Breeding cannabis and continuing a lineage in the seed is not a must for a grower. However, indoor growers that have acquired high-level cultivation skills and mastered the essential techniques can most definitely become interesting breeders. Creating hybrids is also achievable even at a first trial. Most of the cannabis strains that have become legend were created by home growers, even by accident.

While it might not be possible to build your own seed bank from the grow tent in the spare bedroom, small-scale breeding is the next best thing to do. The good news is that you don't need a master's degree in botany to pull this off. Just the normal and ordinary good old-fashioned dope growing experience will suffice. Thus, this is how to go about it.

Same Strain Breeding

Now, this is very easy. Breeding from a same batch or strain can be quite intriguing. Get a male and a female cannabis plant in the same batch, then crossbreed them. For example, if you are familiar with the strain and cropping from the same pack of seeds, you can potentially select a breeding pair to cross.

This is an old-school dope growers' process mostly applied outdoors. Although, breeding from the same batch has potential indoors, provided the original organic seeds are genuine. If so, not only will the resulting progeny be more or less stable but you will have saved cash on seeds for the next crop. Breeding from a reliable batch is a good introduction to cannabis breeding.

Poly hybrids

Ever heard of breeding two different strains to produce outstanding results? Well, a poly hybrid is the crossing of two commercial, unrelated varieties. To begin with, this will not be entirely stable, and won't produce genuine F1 hybrids. Results will undoubtedly be mixed, but poly hybrids are pretty vigorous and winning pheno's can be found in them.

F1 Hybrids

These amazing, genuine F1 Hybrids are the jewels in the crown of the Royal Queen Seeds catalog. The cold, hard truth is that creating fantastically potent, productive and vigorous growing F1 hybrids is a long-term process. It is not child's play, either. It involves lots of processes and procedures. Professional breeders invest years of their lives into breeding projects and select cultivars from hundreds if not thousands of cannabis plants.

F1 hybrids can only be derived from crossing pedigree stabilized or landrace strains. They express genuine hybrid vigor. Unless you're planning a strain-hunting expedition, tracking down heirloom landrace seeds is hard graft. It's probably more convenient to stick with the RQS catalog for awesome hybrids.

Similarly, filial breeding can be complicated. Honestly, it's far too demanding for the first time home breeder. By crossing a pair of F1s (first generation) the resulting progeny is the F2 (second generation). Unfortunately, these seeds will be far less stable and far more difficult to work with than the previous F1 generation.

Careful selective breeding in large numbers is required to succeed with this approach. Often it takes multiple generations of breeding, perhaps until F5 (fifth generation) or even F6 (sixth generation) before the line can be stabilized. That is how much stress the F1 hybrids involve.

We believe this chapter must have opened up your eyes and mind to the terms "flowering" and "breeding" in the cannabis plant. So, instead of surfing the net or a scientific jargon to look for answers you might not find, this handbook will further broaden your horizon in our next chapter.

Chapter Eight

Indoor Growing: Maintaining the Marijuana Plant

Always remember that making mistakes while growing the cannabis plant is very normal. Don't push yourself too hard in order to achieve your desired result. Beginners have a long history of messing things up. There's a learning curve for every activity, and growing marijuana is no different. People who have been growing weed for 20 years are naturally going to be a lot better at it than those who have just decided to start. But, most newbies might be dissuaded from trying it because they fear humiliation or failure.

However, don't let that dissuade you no matter what. Rome wasn't built in one day. Window growing won't cut it most of the time. Despite the fact that the best source of light for any plant is the sun, growing them indoors and using the window as your only light source is a bad way to go.

When it comes to indoor growing, marijuana plants need as much light as you can give them. While it might be cheaper to just try to use the sun, it won't be effective. Buy lights if you're growing indoors. Be prepared. Growing marijuana comes with a lot of vagaries that can leave you feeling overwhelmed. There are also plenty of things that you should just be prepared for.

The plants need water, nutrients, light, and CO_2 (not exactly in that order). But, plants can also be hit with a bug infestation, lack of nutrient quality, and inadequate amounts of CO_2. Make sure you have a contingency plan ready in the event that the plants start to exhibit negative signs.

Use the right fertilizer. Many beginners might just grab any old fertilizer at the garden store. While the plants will grow, they won't thrive like you want them to. Most fertilizers have an NPK ratio conveniently displayed on the bag or another packaging. This isn't just another irrelevant mathematical term, though. It describes the concentration of nitrogen (N), phosphorus (P), and potassium (K) in relation to one another.

For every growth period (excluding flowering), you'll want to use a fertilizer that is higher in nitrogen than anything else. During flowering, the fertilizer should have more phosphorus. Just because the soil is natural doesn't mean it will work. Many new marijuana growers automatically think that any outdoor soil will provide ample nutrients for their plants. In reality, that soil could be nothing more than glorified dirt.

It could also be too acidic or too alkaline and won't even help germinate the seeds properly. When growing outdoors, always infuse the soil with some fertilizer or another potting mix. Also, make sure to test the pH balance to ensure that it's as close to the middle (7.0) as possible.

Be active. As you might have already guessed, growing marijuana is not a passive expenditure of your time. These plants need to be cared for almost like they are your children (however ridiculous that might sound). They have remarkably short lifespans from germination to harvest, but you can't just plant them and hope for the best.

Trim them, prune them, feed them, water them, pamper them, and make sure they're getting enough light, CO_2, and ventilation.

Mistakes To Avoid When Growing Marijuana:

Don't let the plants get rootbound. One thing that many beginners might not know is that marijuana roots grow incredibly fast. When they are in a container, the roots generally line the walls of that container and reach to the bottom.

If the container is too small, they can get rootbound. They will also die. Make sure you (carefully) transplant the plants from smaller containers into larger ones after they've exhibited some accelerated growth (from seedling to vegetative state).

Don't get crazy with pruning. You may have heard that pruning a plant increases growth. You may have also heard that more pruning correlates to further growth. While that can be true, there's no need to prune down an entire marijuana plant. You'll just end up weakening it and potentially killing it if you go too far.

Don't panic. Most of the problems that occur with plants are the result of easily reversible mistakes. For instance, if some of the leaves start to turn yellow and the plant starts to wilt, it could just be lacking in one particular nutrient. Some leaves on the plant will also just die either because of a lack of light or because of natural processes. In general, it's not indicative of a greater problem throughout the plant.

Maintaining the Indoor Cannabis Garden also costs much more money than you might possibly bargain for. It is important to know that almost everything the plant needs is artificial and costs money. From the soil to the pot, the light, the air, and so much more - it all costs money.

Therefore, it is advisable to know what you are getting into before determining whether you are cool with it or not. Maintaining the indoor plant also means checking the plants and the facilities you have in place, so as to be sure if they are perfect and in good shape

Indoor growing also comes with its own fair share of pests. There are lots of pests that would attack the plants if you don't maintain and care for them. Here are some of the deadly pests you should watch out for:

1. Spider Mites
2. Aphids
3. Cabbage Looper
4. Squash Bugs
5. Potato Beetles

Some growers choose to use harsh synthetic chemicals like pesticides and fungicides to keep their plants free of pests. Although these chemical agents might do their job, they also cause great harm to human and environmental health. Residue from these products can even end up in your smoking stash. Thus, the best way to stop these pests from killing your plant is through biological means.

For example, growing some beneficial plants very close to your cannabis plant will help push off the pests of the cannabis plant. Their aroma, scent, and way of growing will help put a stop to the pests eating up your cannabis. There are also some beneficial insects which help eat up the pest that eats up your plant. For example, the spider spins its web around the plants to catch other insects that are pests to your plants. Grow flowers around your weed so they can attract spiders.

Be that as it may, maintaining your indoor plant is paramount for their healthy growth. Focus your attention on how best you can care and maintain your plants. You will be glad about the outcome in the long run.

Chapter Nine

Outdoor Growing: Where and How?

Outdoor growing is different. Unlike the popular indoor growing method which entails growing the cannabis plant in a secluded area of the building (grow room), outdoor growing of the cannabis plant takes courage and guts to pull off successfully. You can't just start sowing cannabis in your background because of the immense land you have in your compound. The outdoor growing of the plant needs planning and strategy.

Most outdoor growers are people with much confidence, self-esteem, and good connections. A single individual cannot just wake up one day and start cultivating cannabis in the open. There are lots of people and processes that must be followed. For example, you will need people to help secure your investment; people to help cater for the plants.

In this line of business or hobby, no one can live the life of complete autarky. However, regardless of how or where you want to begin your outdoor growing, you must first make a decision in your mind about treading this path. Are you well equipped? Are you truly ready? Do you have what it takes? Are all conditions and factors met? If yes, then this chapter will fuel you up with ideas on how to go about it.

Don't get it twisted, outdoor growing and indoor growing of the plant are two sides of the same coin. Simply put, they are different roads that lead to the same destination – cultivating and harvesting good, healthy cannabis plants. They are distinctive, yet geared towards the achievement of the same goal.

The outdoor growing of the cannabis plant is the traditional way of growing cannabis. Before the prohibition placed on the psychoactive herb, growers of the cannabis plant enjoyed complete freedom. Outdoor growing was the order of the day amongst growers of the plant. But currently, that freedom has been cut short.

Outdoor growers now carry out their hobbies in absolute secrecy for the fear of being persecuted, arrested, and jailed. Therefore, if I want to grow cannabis outdoors, I would put it at the back of my mind that this line of business or venture is a very risky one. To this effect, I wouldn't want to jeopardize it in any way whatsoever.

I would find a suitable and conducive place where my cannabis plants would thrive. I would block all possible obstacles that would present themselves as a stumbling block to the success of my outdoor garden.

As an outdoor grower, we often do not feel like we have tied up loose ends, even if we have put in place one of the best security systems one can possibly imagine. We often feel jittery and scared, always feeling like we are being watched with a keen eye. What if we are caught? What if the security is just not enough? These and more are the questions that might keep going through our minds.

But at the end of it all, we realize that we were just being paranoid over nothing. Ease your mind, boost your confidence, and have self-esteem. Remember, it's what you love doing. Nevertheless, all our plans and strategies would be wasted if we didn't find the right place for our outdoor growing.

Finding the Right Place

Like we said above, you just can't wake up one morning and start sowing cannabis anywhere you want. Sounds absurd, isn't it? There are lots of factors to consider when it comes to finding the right place to plant your cannabis, especially with outdoor growing.

1. **Tightly Secured Environment:** Security is one of the key ingredients of any establishment. Be it an investment, a parastatal or, as in this case, an outdoor growing farm. There should be laid down processes, structures, and arrangements which provide top-notch security over the garden. For example, a sudden surge that suddenly occurs in the amount of electric current used within a short time would definitely attract unwanted attention from the relevant authorities.

Also, an increase in the water bill can charge up interest from other authorities. How would you settle those issues? What necessary arrangements have you fashioned to tackle this breach? These and more are relevant questions you should ask yourself.

Finding the right place for outdoor growing of your cannabis plant is no small task. Ensuring maximum security is of utmost importance. In other words, the site must be tightly secured. It must be away from the public eye.

2. **Large Area of Land:** An outdoor growing area needs to be spacious. The grow space must be large enough to accommodate all seedlings. Unlike with indoor growing where seedlings are housed inside a building with air, water, and light provided artificially to ensure good yields, outdoor growing is just natural.

   Natural soil, natural air, and natural light directly from the sun; all these can be achievable with a large area of land. It creates space for the seedlings to spread out and reach their peak. To this effect, the seedlings won't be competing for nutrients, air, water, and light to survive.

3. **Good Climate:** According to research, a good climate condition is vital for the growth of a healthy cannabis plant. These climate conditions which are not too hot or too cold provide the plant with a perfect means of reaching its best height. A study conducted in this field has shown that a climate condition which can sustain the tomato plant would also sustain the cannabis plant.

   According to the study, tomatoes and weed were planted alongside each other and the results were very encouraging. They both excelled in that particular weather condition. Thus, wherever you can grow tomatoes, you can grow cannabis, too. In choosing where to start your outdoor cannabis farm, putting the climate condition into consideration is vital. You don't want to grow your plant in a harsh environment.

4. **Good Soil:** First thing first, looking for a place with very fertile and healthy soil to grow your cannabis plant is of utmost importance. And if the soil is not treated before planting, we would advise you do so before starting your cultivation.

   Additionally, you want to remove as many weeds as possible. These weeds can compete with your plants for nutrients if left alone, thereby making them grow weak, unhealthy, and withered. All the above-listed points are vital to finding the right place for your outdoor growing area.

   The security of the place must be guaranteed. Even when the authorities pull unnecessary and unexpected stunts, there should be arrangements laid down to checkmate these stunts. A wide area of land, a good climate condition, and a nutritious soil should also be the key factors towards choosing where to sow your cannabis seeds outdoors.

How to Sow Cannabis Outdoors

The outdoor growing of the cannabis plant can be quite fun, especially if it involves more than one person. A lot of cannabis smokers out there would easily agree with us that outdoor cannabis is the best of them all. Many of them even went ahead to describe the outdoor cannabis as the perfect weed, with a scintillating scent they just can't resist.

This should tell you a lot about outdoor growing cannabis. With the right place to sow, the right weather conditions, and perfect security, there is nothing stopping you from burying the seed in the ground. Well, except one question; how do you go about it?

You have a plot of land already? That is just perfect. Start by treating the ground. It is important to note that untreated ground ends up producing very low yield or harvest. It is also prone to pests and small predators ready to reduce your hard work to nothing. Therefore, for much better yields and harvests, we would recommend you start treating your land henceforth.

Remove all disturbing weeds you can find. Weeds can be very stubborn most times. To this effect, you shouldn't relent even after removing them totally. They have the tendency to come back even stronger and thicker. Make sure you don't get lazy when it comes to weeding out your farm. Employ any technique you can think of, so long you don't end up hurting your plant.

Don't leave the weeds around carelessly for people to see. This may attract unwanted attention and attractions from passers-by. Tie them in a sack if you must or dispose of them far away from the farm. Afterward, sowing your cannabis seeds should come next. A lot of people end up making the mistake of sowing the seeds too deep.

If you are one of those people, then please stop. Here is why! When the seeds go much deeper than expected, it ends up taking more time to sprout or they even might end up not growing at all. Adding pre-made soil with an NPK value won't be a bad idea at all. Getting the pre-made soil should be easy as they can be found at the nearest store close to you.

This is where continuous weeding comes into play. Week in week out, weed out any form of obstruction till your seeds germinate into seedlings. Don't forget to sprinkle water every now and then, if you have to. Growing cannabis outdoors is as easy as that.

Like we discussed earlier, there is nothing new about cannabis planting, however, knowing how to plant the seeds appropriately, knowing where to plant, and ensuring some safety measures are taken towards the protection of your cannabis plant is paramount. The next chapter will delve into the outdoor planting techniques. You don't want to miss it!

Chapter Ten

Outdoor Growing: Planting and Planting Techniques

Outdoor growing differs from the indoor method of growing as well as in its planting techniques and styles. A grower who wants to adopt the outdoor growing method would not use the indoor growing techniques, and vice versa. The end result is bound to be disappointing, disappointing, and degrading in the long run.

Where indoor growing of the cannabis plant uses artificial lights and air, outdoor growing of the plant doesn't really need the artificial setup. Imagine installing the light and air conditioning equipment outside. Crazy, isn't it? That is one of the reasons why lots of expert growers prefer using the outdoor growing to the indoor counterpart.

Growing cannabis outdoors is a lot cheaper than indoors. When it rains, it's free and natural water to your plants. When the sun shines, it's free of charge - without cost. Even the air the plants take in is totally free. That way, you spend less when compared to Indoor growing. However, you can decide to use a grow pot outdoors.

Grow pots come with their advantages and disadvantages. Inasmuch as they can be moved about easily while changing positions and postures of the plants, they are very advantageous. However, the negative part of the movable outdoor grow pot is that they sometimes deny the plant good light, the great nutrients they could get when planted on an unlimited space, and the good air they would get when outdoors.

To get the best out of your cannabis plants, one must learn the necessary planting techniques and know exactly how to apply these techniques at the right times. As a beginner, these esteemed techniques, skills, or styles might be news to you. If you find yourself in this category, don't feel bad. There is always a first time for everything. Even the expert growers out there started out as beginners.

Thus, with time, as you would expand your knowledge and learn these unique techniques, you would find it very easy and entertaining as you apply them. To that effect, this chapter would go a long way in opening your eyes to the advanced techniques you can use to upgrade and increase the yields or harvests of your plant.

These techniques come with both indoor and outdoor growing but are more synonymous to outdoor growing. This chapter will broaden your horizon on the positive side of these techniques before applying them. Growing is very easy. Sometimes all we need to do is to bury a seed and pretend to forget it even exists.

Like we said earlier, planting techniques are very much advanced in their procedures and processes. They go beyond the normal procedure we discussed in the previous chapter. Exercising these techniques on your farm would not only result in a better output or harvest but also better cannabis efficacy and genetics.

This would go a long way in ensuring and improving the qualities of the plant's bud cultivation. Get a handbook, if you must. Ask a friend or two to teach you how to apply them, if it comes to that.

Planting Techniques

All planting techniques are unique. But the good news is you can use two or more of them at the same time. They complement one another.

1. **Thinning:** Thinning is mostly synonymous with outdoor growers. Imagine growing your weed from inception in an orderly and uniform manner, only for them to start growing shabbily and aggressively or even out of line. What would you have done? One would notice that some of the plants are growing faster than others, racing toward the light.

    In this case, adjusting the light to suit other plants would be the best thing to do, or better still, moving and switching the plants is also great. Spacing also comes to play here. However, if even after adjusting the light and the space between the plants, your plants still grow abnormally, thinning is the next thing you should consider.

The racing plants tend to draw closer toward the light at the expense of the others. Therefore, other plants that don't get much light end up growing slowly, shorter, and abnormally. Thinning these racy plants becomes paramount so as to give room for other plants to absorb the sunlight.

Cut them down, trim them, pare down, or clip them to the normal size of the deprived plants. Trust me, when all plants are of uniform height, they will thrive together in harmony. The end result will be bumper harvests and surplus yields.

2. **Cloning:** Ever heard of the term cloning? I'm sure what is going through your mind right now is the duplication and replication of a phenomenon, person, or something. Well, you are right. However, cannabis plant cloning doesn't mean the plant would be taken to a lab and be put through lots of chemical processes just to get a replica. It is simply the planting of unwanted plants you removed earlier in your farm.

   Do you remember those unwanted parts of your plants you thinned out? Yes, those parts! Why throw them away when they can be planted again? The act of planting those unwanted plants is called cloning. A lot of growers end up throwing the thinned plants away. But rest assured, by the time you are done with thinning your plant, there will be more than enough space for you to clone. To this effect, your harvest will increase drastically.

3. **Plant Shifting:** This type of growing technique is applicable to mostly outdoor pot growers. Like we discussed earlier, some plants have the tendency to grow faster than the rest. Thus, growing towards the light. This is a hindrance to other plants as they may end up not getting enough light.

   Thinning would be the best solution but some growers don't like thinning. They believe that thinning slows down their cannabis growing process. Be that as it may, switching the plants becomes paramount. Interchange the position of the faster-growing plants with the slow ones. This should regulate the growth and size of the plants. And trust me, the slow-growing plant will catch up in no time.

Notwithstanding, if your plants cannot be switched from one place to another, try other means. For instance, in the hydroponics style of planting cannabis might be very difficult to move from one place to another. Therefore, using another means of moving the plants like tying a rope to the fast-growing plants in a straight form also works just fine.

For outdoor cannabis plants, you can use sticks, bamboos, and stakes with a very long, thick thread to hold the plants in an upright position. Exercise the planting technique when the need arises and watch your harvest quadruple.

4. **Pruning:** Do you often wonder how some of your friend's cannabis plants end up getting more than one flower at the top colas? I'm sure you must be thinking your plant is growing abnormally. That is not true. As a matter of fact, your friends exercised a planting technique which they obviously didn't want to let you in on.

   This technique is called pruning. At the 3rd or 4th week of vegetation growth, the top should be pruned off from the stem, which will give the main stem an opportunity to sprout in two or more directions. It no longer takes the normal straight shape but a remarkable "V" shape.

   Nevertheless, this technique does not always work out all the time. The plant's response to the technique depends solely on its strain and surroundings. Topping off a plant sometimes produces more than five top colas. Some Cannabis strains also reach their maximum bud production only if they are topped off. In other words, pruning leads to more bud production and more bud production equals bumper harvest.

With everything we've discussed above, increasing your yields and harvest shouldn't be hard to pull off. From the life cycle of the marijuana plant to its germination, indoor and outdoor growing, factors that may influence growing marijuana, and so much more.

All these should give you more than enough ideas on how to go about establishing your own private cannabis garden. The questions now remain: Do I really have what it takes? Can I actually pull this off? Growing marijuana can be very addictive. Far more addictive than its content (Note: Marijuana does not contain any addictive content but the love of growing it can be very addictive).

To many people, it is the hobby in which they derive utmost pleasure and contentment. It is far more than just a business venture even for those who commercialize growing cannabis. Growing marijuana is rewarding in the long run.

To this effect, using planting techniques on your cannabis plant would not only increase yields and harvest, but it will also ensure that cannabis growing outdoors is much more rewarding than its indoor counterpart. These advanced techniques would give order and uniformity to your plants. Quite rewarding, isn't it?

These diverse techniques are different paths that lead to one destination – a great harvest. Therefore, we would advise you to be like your friends. Apply these techniques where they are meant to be applied and of course, appropriately. The joy of all cannabis growers, expert and beginner, is to see their plants blossom. Don't be left out.

Chapter Eleven

Outdoor Growing: Care of the Growing Plants

Jamie was very passionate about his cannabis plants. Every morning, he would weed, water, and check his cannabis plants just to be completely sure they were in good shape. As a key follower of the Growing Marijuana Handbook, he always made sure he followed the handbook to the letter. Thus, he continued the processes as outlined by the book.

This process continued until Jamie traveled for a business trip out of town for two weeks. Before leaving, he had checked with his friend Sam to help care for his precious plants. Sam wasn't Jamie, thus he would forget to check the plants frequently. Day in day out, the plants continued to wither away due to Sam's negligence.

They continued dying off with pests and smaller predators perching on the leaves, branches, buds, stems, flowers, and trims. This was solely because Sam didn't pay much attention to the care of his friend's precious plants.

Now let's imagine you were Jamie. How would you feel after coming back from a long distance journey, tired and weary, only to see your precious plants reduced to rubbish? Unimaginable, right? To this effect, we would advise you care for your plants yourself. No one takes care of a property more than the owner. Leave no one in charge!

But if you must, then always check in with the person frequently. Knowing how to grow cannabis, even in some spectacular way, is not enough. It is important to note that even as you become an expert grower, knowing all the techniques and strategies towards growing healthy cannabis, if you don't take care of your plant, it's going to be a wasted effort. A healthy cannabis plant can be only produced via care and attention given to it from the seedling stage to the harvesting stage.

Care of the plant means you nurture, nourish, and sustain the quality, quantity, and caliber of the cannabis species you are growing. Sometimes, even the sight of our outdoor cannabis farm scares us. With the large area of land, we begin to imagine where we would even start from. If it's weeding, then weed early and weed frequently. That way, the weeds won't get to encroach on the plants.

Care of the plants takes various procedures and processes. They vary from the kind of growth you adopt. In other words, indoor growing care is totally unique from that of outdoor growing. This can be because of their indoor and outdoor features. For example, due to the large space provided by Mother Nature, the outdoor cannabis plant care includes weeding out the large chunk of unwanted and irrelevant plants.

However, these unwanted plants are not present indoors as they can be hardly seen. This is because of the nature of planting indoor plants. Indoor plants also grow in a pot, box, and bucket. This gives them limited space to get entangled and spread their leaves and branches as far as they want

Care of the Growing Plants

1. **Weeding:** According to expert growers, the best and most convenient way to weed your cannabis plants is by hand. Also, regular growers tend to use killer weeds or weed killers, as the case may be, around their Cannabis plant. It is believed that these weed killers may have a negative effect on your cannabis plant if it's not applied carefully.

   However, the overall benefits of these weed killers on your plants cannot be overemphasized. We would recommend you first try it on your clones in order to test the effect it would have on the cannabis plant firsthand. Little wonder why weeding by hand is the preferable choice to the expert growers out there. It is, no doubt, the most acceptable method of weeding to date.

   Mother Nature often starts sprouting annoying weeds all over your cannabis plant farm. They even get to a point where they start getting intertwined with your plants, sucking and tapping off nutrients from the plant. You definitely can't remove the weed aggressively. You might end up affecting the cannabis plant.

Thus, a careful weeding with your hands is the solution to that problem. A weeded area of plant ensures that the plants don't get to fight or compete for any nutrients, sunlight, air, and water. It paves way for the plants to grow healthy and reach their peak. A weeded cannabis plant benefits the plants immensely as they spread out without any form of obstruction or competition.

I guess you must be seeing the issue of weeding as a very demanding one. Well, not to worry. According to expert growers, weeding your growing plants need only be frequent during its first 2-3 weeks. To this effect, weeding is advised to be carried out every week. But after this stage, it can be done once a month until your plant reaches the harvest stage.

Nevertheless, if your area of land is densely populated with weeds, then weeding by hand or the use of weed killers becomes irrelevant and a total waste of time. Imagine weeding a hectare of land with your hands; absurd, isn't it? This is where ground covers come into play.

2. **Ground Cover:** Ground cover, as the name implies, is a kind of cover which is spread on the ground with different sizes of holes bored into it. These holes provide space for sowing your cannabis seeds. Ground covers can be anything that can be spread on the ground to stop the intrusion of a common weed. They can be plastic, bin liners, or even sheets of paper. It will definitely keep the weed down.

3. **Watering:** When it comes to outdoor growing of the cannabis plant, Mother Nature should be in charge of watering the plants with her rain. It is her job as all your water should come from the sky. Hitherto, if the season changes from a raining season to a dry one, watering the plants yourself becomes paramount.

The kind of watering method or equipment you will use solely rests on the size of your farm. You certainly won't want to use a watering can to water a hectare of land. For a small area of land or even outdoor pots, you can use the watering can, a cup of water or even put water in a sack if the farm is a little bit far.

For a large area of land, expert growers make use of a sprinkler system. This is very fast and can reach other plants that are far apart. Like we said earlier, the size of your farm determines everything, which includes how much water to be sprinkled.

Some cannabis plants make use of more than one gallon of water per day, while some just need a cup of water. Knowing when the plants need water is paramount. The cannabis plant has the tendency to hold pockets of water below the surface. But how does one begin to tell which plant needs water and which one doesn't?

It's simple. Each cannabis plant tends to wilt badly when it is in dire need of water. Therefore, whenever you see your plants getting wilted, you now know the reason. Another way of knowing this is to keep a foot-deep hole beside your plant. Be very careful when doing this so as not to damage any vital roots.

Then proceed by feeling the soil with your hands. How moist is it? Is it completely dry? These are questions you would get your answers to, after feeling the soil.

4. **Safe from Predator and Pest:** There is absolutely no healthy plant without its fair share of this problem. Predators and pests eat through your plants until they die off. Without enough care, they will easily find a way into your plants, thereby reducing the quality, quantity, and caliber of your plants.

To salvage your plants from this path of destruction, one must do the needful. This may include using cut-throat measures and means to achieve your objectives. Expert growers have agreed that the number-one defense system one can use in eradicating these small predators is the cat. No small predator dares to enter the house of the cat, except for the cunning and daring ones. That will solve your problem on the predators.

For pest control, please ensure you use only pesticides that are labeled with "For Food Product Use." If you don't see this label on the pesticides, then do not use it on your plants. Follow the instructions and manuals of the breeder to the letter. You wouldn't want to get sick from taking in harmful cannabis.

Some of the pests and predators you might encounter include:

- Woodchucks: They bite off the stems, chunk after chunk, until they eventually fall off the plants. Use predator urine or construct a meshwork fence around the plants.
- Powder Bugs: They destroy the plants by constantly laying their eggs in the bud and stem. This will kill off your plants within the shortest possible time. Spray a Pyrethrum-based insecticide on your plants. This keeps them away.
- Groundhogs: They continuously feed on your plant until it dies off. Get some dry chloride and apply it around their holes. This should keep them away.
- Wilt Fungus: As the name implies, it wilts your precious plants completely. Gently apply the fungicide. You can find it in a store nearby. It works like magic.
- White Flies: They are very deadly. Within days, they can destroy your plants, if given the chance. Ensure you get the original safer's soap. It is very potent in killing off the whitefly. It can be found at any grow shops close by.

Everything needs care and attention to be done or turn out perfectly. Else, they might fall through in the process. The same thing applies to the cannabis plant. Remember, it is your passion and hobby. Treat your plants with love, care, and attention - and watch them blossom. A lot of obstacles, factors, hindrance, and stumbling blocks could create problems for you along the line.

Potential problems abound, like pest and small predator issues, lack of nutrients in the soil, racing plants, abnormal growth, and so much more. It is your job to ensure each problem is taken care of adequately. Consult a friend if you must; read a book or surf the net if you have to. Lastly, no one takes care of the plant better than the grower.

Chapter Twelve

Outdoor Growing: Nutrients and Environmental Factors

In the previous chapters, we basically surfed through the processes and procedures of growing a cannabis plant. Most of the work and praises are heaped on these processes, but we must not neglect the important roles nutrients and environmental factors play in the healthy and rapid growth of the cannabis plant.

The soil contains major nutrients (Nitrogen, Phosphorus, and Potassium) which nurture the plant from its germination stage. These nutrients, which are also known as NPK, can either come with the soil or as a fertilizer. These nutrients come in equal proportions (20:20:20). The remaining 40 percent contains other elements and contents that make up the soil.

A soil without nutrients can be said to be a train without an engine. There will be no way the train will move. This same instance applies to the soil. There is no way the cannabis seeds will germinate with a soil that lacks nutrients. Please, always remember that each plant has a particular nutrient it wants more than the rest.

The cannabis plant wants more of the N (Nitrogen) and the normal amount of the P (Phosphorus) and K (Potassium) during vegetative growth. Additionally, during the flowering stage, use nutrients that are higher in the P (Phosphorus). Knowing all of this is of utmost importance for the growth of your cannabis plant.

The sun, air, and water also go a long way in nourishing your plants to a large extent. But with the addition of nutrients, it will only get better. Here are top nutrients you can add to your outdoor Cannabis farm;

1. **General Hydroponics Organic Go Box:** Sourced from botanical extracts and natural minerals, these organic nutrients will boost your plants with the necessary things needed for a healthy take off.

These nutrients will provide additional support to the already existing nutrients. It supports the roots, makes the leaves look brighter and greener, and helps out in the different phases of the plant.

2. **Dyna-Gro8 oz Dyna GB8OZ Liquid Grow and Bloom:** This is a very good choice of products for all weed growers, both indoor and outdoor. You will find this product incredibly easy and convenient to use ahead of other products in this range. It provides the plant with all the necessary nutrients right from the beginning.

    The Dyna-Gro Liquid Grow (8oz) should be used first to boost the growth and size of the plants, while the Dyna-Gro Liquid Bloom (8oz) should be used during the flowering stage. This will help bloom the flowers and buds into becoming much more effective, productive, and of great quality. It is perfect for beginners.

3. **General Hydroponic 1 Quart GLCMBX009 Calimagic & Flora GRO, Micro & Bloom Combo:** This is just the perfect fertilizer you need for correcting soil with a deficiency in potassium. It also boosts the rapid growth of a plant by giving it the necessary food for that. This particular nutrient has lots of important functions.

4. **Espoma TR44-Pound Tree-Tone 6-3-2 Plant Food:** There is no better natural food you can feed your plant aside this fertilizer. Tired of chemical-based fertilizers? Then try this natural fertilizer from Espoma. It contains a massive embodiment of over 13 nutrients which will nourish your weed. It can be used from the initial stage of germination until harvest.

5. **Clonex Rooting Gel:** Cloning is a technique used in planting thinned out plants. With Clonex, this thinned plants that are cloned will grow faster and better. The rooting gel works like magic by repairing, protecting, and stimulating the root growth of your cloned plants for a much better plant in the long run.

Environmental Factors

The environment is everything when it comes to planting cannabis. If you plant your cannabis in an environment that is not conducive, the results will most definitely be bad. You can't expect your plants to grow perfectly in bad environmental conditions. These factors can deter, delay or even lead to damage of the plants in the long run. The environmental factors are complex. They are difficult to control and most times just simply out of our control.

1. **Land Management Practices:** This is an effective factor that influences the growth of outdoor marijuana. Land clearing, habitat fragmentation, and over-grazing help the growth of weed by providing the clones with a new area of land and ensure the weed doesn't have a strong competition with native vegetation.

   Land management practices are activities which are carried out to improve the quality nutrients and fertility of the land. To this effect, the marijuana plant gets the necessary nutrients it needs from the soil.

2. **Natural Disasters:** As we discussed earlier, growing outdoor weed comes with lots of factors. Natural disasters are events that are beyond our control but end up leaving our plants in ruins. Natural disasters like cyclones flood and drought can cause problems to the plants.

   The cyclone can cause strong turbulent wind which will destroy the plants far beyond measure. It disperses the soil and even roots out the plants. The floods also wash away the plants. Floods are equally disastrous to the plants as they fall plants, uproot them, and damage them. Drought, on the other hand, dries up the whole land. It dries every water and moisture of the soil to the extent that the plants dry up easily and completely.

   However, these natural disasters can also create the perfect conditions for the plant to thrive. With other native vegetation drying off, being washed away, or dispersed by the turbulent wind, weed plants get the avenue to enjoy all the nutrients in the land all to itself. To this effect, the plant gets little or no competition from the native vegetation.

3. **Fire:** It is important to note that the cannabis plant is highly flammable. One of the primary methods of taking in the cannabis plant is by smoking, thus be sure to keep fire far away from your farm. As a matter of fact, fire and weed hold a very complex relationship as it is very deadly and suppresses the rapid growth of the cannabis plant.

   However, there is certain weed that often benefits from fire either directly or indirectly. Fire helps reduce the rate of competition by burning out other native plants in the farm. This would give the weed plant the necessary space it needs to thrive.

4. **Climate Change:** Climate change is not suitable for plants in general. But this is different in the case of marijuana. Cannabis' aggressive way of growing gives it an advantage over other native vegetation. It thrives even in the harshest of conditions. Climate change impact is most severe on the plants.

   Even though the Cannabis plant tends to grow in this condition, it won't reach its best in terms of yield and harvest. Climate change is often caused by variations, biotic processes, and human activities. In other words, it is also called global warming.

   Some species or strains of weed that adapt to this severe climate condition end up with many advantages, which include more than enough space to expand their growth, little or no competition, and surplus nutrients.

Like the human body, even the cannabis plant needs nutrients to flourish. In a case where the soil lacks enough nutrients to nourish the plant, we would suggest you buy a fertilizer containing the necessary nutrients it needs to thrive. Additionally, no plant can survive in a severe environmental condition. Be that as it may, our next chapter will definitely be much more interesting. It promises to enlighten as well as familiarize you with the beautiful stage of the cannabis plant – Flower stage. Let's read through, shall we?

Chapter Thirteen

Outdoor Growing: Flowering and Breeding

Flowers are synonymous to most plants you would find out there. They blossom so brightly and beautifully, signaling the start of harvest. To this effect, when you see your cannabis plant sprouting flowers, then preparing for harvest is the next thing to do. However, before the cannabis plant reaches this far, it must have passed through lots of rigorous processes, techniques, and care.

The flowering stage of the cannabis life cycle is the most sensitive stage of the plant. This is when the plants need more than enough care and attention. Like we discussed in the previous chapters, the flowers are formed as a result of the male pollen sack being too ripe, thereby bursting all over.

This burst pollen sack disperses pollen all over the female plants. In turn, the female brings out white hair which would be visible at the internodes and top cola. In other words, these hairs are called "pistils". They start getting longer, thicker, and have more curl. They are mostly sticky and covered in resins.

According to expert growers, the cannabis plant produces these resins to capture pollen. Thus, if it fails to capture enough pollen, the plant (female) would produce more resins all over their body just to capture more of the male pollen when they are falling off.

The flowering period also will witness the plant filling out more than usual. The spaces between the branches and stems will be filled out appropriately, giving the cannabis plant a Christmas tree shape. There will be bigger leaves filling up each branch. The lower fan leaves will draw more of the sunlight. This will give the plant a better floral and leaf development.

To notice the peak of the flowering phase, the pistils of the female cannabis plant, which is located on the flower top, will swell. This swelling will make the pistil change color gradually. In most cases, it will change from white to brown. In other cases you will see it change from all white color to an orange tint, after which harvest is ready.

It is important to note that each strain or species of the cannabis plant comes with its own flowering moment and times. Each strain and species may also come with a whole different color when they reach their flowering peak.

Flushing the Cannabis Plant

We would advise you to always go through with this procedure whenever you are at the flowering stage or a few weeks before harvest. When you flush your plants, you basically put an end to the series of nutrients and fertilizers you feed them with. Only water that contains a neutral PH would be administered to the plants within these last few weeks.

How does flushing work? It's pretty simple. What flushing does is push out the salts and minerals in the soil. This will help give the plant a much better taste and scent, unlike the unpleasant, bad, and chemical-like taste and scent you could experience otherwise.

Also, during the flowering stage, you can choose to boost the quantity, quality, and caliber of the buds your plant's sprout. This is possible through the use of bloom fertilizers (as discussed in our previous chapter) and nutrients. Give the plants these blooming fertilizers during the first and second week of flowering.

This would make them grow taller and blossom more. The size of the plants you have in mind determines the extent of the fertilizers you should use. In a situation where you want your plant to be of small size with bigger buds, the bloom fertilizer is your best bet.

When the plant reaches the flowering stage, the production of THC increases drastically. This is very good for your plant if you intend to smoke it. The sticky nature of the THC gives the resin room for it to hold tight to the plant. THC is also used in permanent solutions to bug problems. It is also the flavor, the psychoactive substance, and the high that makes smoking cannabis intriguing.

Outdoor cannabis flowering stage is less stressful than with indoor growing. The outdoor nature gives the plant an edge as the earth tilts. Getting the necessary darkness for the flowering of the cannabis plant can be quite easy with the help of nature.

Breeding the Outdoor Cannabis

Breeding is a very important process of planting cannabis. Cannabis plants can be grown for two purposes which are:

1. As a means of reproduction
2. As an independent plant

Breeding is very easy when compared with other processes, techniques, and procedures that come with growing the cannabis plant. There are often strains that possess a great quality of plants and we would want to replicate them and those desired qualities. Breeding is the only way you can achieve those qualities.

If two plants of the same strain are bred, they end up producing seeds which contain almost all the same traits of the parent plants. It is important to note that the seeds might not contain all the traits of the parent plants. Both the offspring and parent will vary on either their level of potency, their color, the flavor in their taste and scent, etc.

This seed-making procedure is very easy. It is the process in which the male cannabis plant's pollen sac gets ripe and as a result, gets burst all over the female plants. This will pollinate the female plants and in the long run, the buds will hold seeds.

This pollen can be collected in cases where the male plant produces more than enough. They can be stored or preserved in film canisters for a period of 18 months. They can also be kept in freezers, pending the next harvest. When the male plant opens up their pods, it is best for you to carefully and gently collect this pollen. After getting enough, endeavor to shake off the remaining pollen on the female plants for pollination.

How to Breed Pure Strain

A lot of expert growers prefer replicating some seeds they bought with outstanding qualities instead of churning up strains to get a crossbreed or hybrid. To begin with, get the preferred seeds you would want to breed without mixing it up with other strains. It should be strictly similar strains. Afterward, sow the seeds and watch them grow. Be sure to allow your farm to contain just the strain you want. We would highly suggest you don't introduce or bring up new or foreign strains later.

Allow the male plants and the female plants to flourish well together. During the pollination phase, allow the male plant with the same strain as the female to pollinate. The end result will produce seeds of the same qualities. However, the qualities might not be a hundred percent replica of the parents.

How to Breed a Hybrid

Breeders are mostly fond of making new strains from the cannabis family. They often see planting and breeding of the cannabis plant as a way to showcase their love and passion for the hobby. This hobby has made them come up with a new form of weed, which is known to be called a hybrid. The hybrid is a crossbreed of two or more strains in producing an offspring with outstanding qualities. These qualities would be a mix of the parent qualities.

For instance, if a local dog mates with the foreign one, the offspring qualities would be the combination of both dogs. The offspring might pick the fine skin of the foreign dog, the long oral cavity of the local dog, the fine body build of the foreign dog, fluffy hair of the foreign dog, and so much more. This same thing applies to the cannabis plant.

The male plant from a different strain would be grown alongside a female plant of another strain. After pollination, you would notice that the seeds would be different. They would contain the mix qualities of both strains. However, they may also contain qualities the parent strains do not possess.

Thus, breeding may seem pretty easy, but trust me, it has its own fair share of complexity. If you are not knowledgeable about the procedure, your plant might start producing the wrong seeds. Be sure to familiarize yourself with the basics of genetics. Have an idea of how the genes work, which plant contains the homozygous, which is heterozygous, the phenotype of the plants, as well as the genotype.

These and more make up the important qualities and features of a strain. You can't just start crossbreeding plants of different strains without checking to see if they are compatible. A lot of beginners make this mistake, after which they would start searching the internet for answers to their problems.

Flowers and Breeding of the cannabis plant are two different things entirely. Where the former depicts the start of harvest, the latter entails the reproduction of similar strains or a whole new strain entirely. We are sure this chapter has helped you understand the importance and sensitivity of the flowering stage. Thus, we hope the next chapter would hold the same thrill. Let's turn the page over, shall we?

Chapters Fourteen

Harvesting the Marijuana Plants

Harvest is the end result of every plant. It is the long-lasting result all cannabis growers anticipate with much excitement. As a matter of fact, a certain thrill comes with the harvest. Knowing full-well that your plants of many weeks are to be harvested for the same purpose (cannabis intake) that motivated you in growing them can be quite intriguing.

Harvesting your cannabis plant marks the end of its life cycle. But before you start getting all excited, are you ready to face the stench smell it will produce during this stage? If no, then we would suggest you keep this in mind before proceeding to harvest your plants.

Most breeders attach a manual to their seeds. This manual also contains instructions to follow when harvest comes. Beginners often don't know when the plants have reached the harvest point. Sometimes, the flowering times which are written on the manuals are quite different from reality. The plant may present a whole different situation where the flowering times might come ahead or behind the breeder's timing.

The plants might start producing flowers a few weeks earlier than what was written in the manuals. To this effect, it is important to know when the plant reaches their harvest period through their features and appearance. Also, these changing features differ from each strain.

Indica Harvest

The Indica harvest is quite similar to the Sativa harvest. The Indica plant, which we discussed to be a 1-4 feet tall, should be cut down from the bottom, after which, you hang them in an upside-down posture. Remove as many of the fan leaves as you can. Do the same for the secondary leaves, but don't dump them in the same pile as the fan leaves. You can also clip out the trim of the plant. The trim makes up great quality weed. They are the leaves that have resins all over them.

Be that as it may, you now have more than two kinds of weed available at your disposal. Firstly, the fan leaves, which contain the lowest amount of THC. The secondary leaves, on the other hand, which possesses higher THC than the fan leaves. They are very intriguing and come with many flavors. Lastly, the trim, which happens to be the best leaves you can get in the Indica species. They are dipped in resins, which makes them much more potent than the other two leaves.

Sativa Harvest

Unlike the Indica harvest that doesn't require much work, the Sativa harvest is very strenuous and stressful as it contains lots of leaves and branches. The Sativa plant is over 10 feet tall. This is why outdoor growers like growing this plant more than its counterpart. Most times, they contain over 20 oz of buds. Now just imagine the heavy workload right in front of you!

First of all, we would advise you spread a sheet, or a canvas spread as the case may be, on the ground, so as to prevent the buds or leaves from falling to the ground. Just like the Indica plant, it must be cut down from the bottom and laid down on the canvas spread or sheet. Afterward, wrap them up nicely and gently with the canvas spread so they can be transported safely to where they would be harvested.

Just like the Indica plants, hang them upside down before carrying out the harvest process. Hanging them up without cutting off branches would present you with problems when clipping off the leaves as they would be too bushy and filled up.

Break off the branches and hang them up separately. After which you should proceed with clipping off the leaves as we have discussed in the Indica harvest section. Please note that too much direct light decreases the quantities of the THC. Where ever you plan on harvesting this plant must be relatively dark.

Harvest Time

Harvesting cannabis at the right time is vital. It would allow the bud to reach its maximum potential and growth, thereby making the THC much more potent. This is why many smokers end up rejecting some weed by merely looking at its appearance.

Additionally, choosing the right time to harvest your plants is influenced by a number of factors. According to expert growers, they believe harvest timing should be strictly a matter of personal preference. Sativa and Indica bloom phases differ. Thus, the harvest of the species varies in time. Here are some of the features to watch out for to tell the right timing;

Color Change

To begin with, there are a couple of ways to tell when it's time to harvest your cannabis buds. Perhaps the most reliable is to examine the color of the trichomes. These resin-bearing glands are considered the best standard of measurement as they are more consistent in their results than other recognition methods. In order to do this, you will need a magnifying glass—trichomes are quite tiny.

It is important to know that trichomes go through three consecutive color states. These are clear, cloudy, and amber. The best time to harvest is when half of the trichomes are amber, and half is clear or cloudy. This color disparity is due to the uppermost buds ripening earlier than the ones at the bottom. In any case, you don't want to wait for all trichomes to turn amber, as this generally leads to a decrease in THC.

Leaves Turn Yellow

This is one of the glaring features you would see in the plant. The yellow leaves are a good sign that your cannabis plant is ready to harvest. You can flush your plant when its large fan leaves start to turn yellow or when they fall off by themselves. But bear in mind that the falling of leaves is unlikely to happen when you use fertilizers.

Pistils

If you've got a photoperiod cannabis plant, then checking its pistils and stigmas is a useful way to gauge whether it's ready for the chop. You can usually assume that a plant is ready to harvest when about half of the pistils are brown. However, we would still recommend the trichome method as it is much more reliable.

Curling Leaves

This is one important clue to watch out for in the cannabis plant. The drying and curling of leaves is another sign that your cannabis plant is probably ready to be harvested. This happens because your cannabis takes less water as it nears its final phase of life. But be sure to confirm this method in conjunction with others, as there are pests and diseases that can cause dry and curling leaves.

Breeder Schedule

Like we said earlier, the breeder's timing might be wrong. But it's still a good way of timing your harvest. It is important to at least consider the harvest schedule provided by your seed source. This is usually found on the seed packaging. This schedule is the approximate number of days/weeks it will take for your cannabis seed to grow into a mature plant. This schedule, however, fluctuates based on growing conditions such as environment, water, and heat.

That is all you need to know to know about harvesting the Cannabis plant. What you do with your plant after that is totally up to you. Harvesting the plant comes with excitement and thrill. It is the final reward for the hard work you put in over the last couple of months. Feel free, enjoy your cannabis, it's the fruit of your labor. However, aside from smoking your harvest with friends, are there other things you might be able to use your cannabis plant for?

If you are interested in finding answers to that question, then we suggest you read on to the next chapter. You will surely be amazed at what you can do with your harvest.

Chapter Fifteen

After Harvesting, What's Next?!

I'm sure you must be thinking after harvesting your cannabis, smoking it with friends is the next thing to do. Well, you might be right, but you can't just start smoking wet leaves and buds. There are steps that need to be followed and procedures that need to be set in motion before smoking the plant can become a reality.

The process of growing cannabis does not stop at harvest time. Properly drying and curing your fresh cannabis stash is paramount to prevent mold contamination from taking place. These procedures will also result in buds that taste better and offer a superior high.

Drying and curing your Cannabis plant is a vital part of the final consumption process. It singles out your Cannabis plant from other organisms that may perch on the leaves, buds, and branches. When the leaves are clipped off during harvest, the leaves are left for a while. Please always remember not to mix the leaves up. The trimmed should be separated from the secondary leaves and the secondary leaves are separated from the fall leaves.

When setting out with the intention to grow cannabis, it's easy to see why harvesting handfuls of treasured buds is considered the pinnacle of the experience. All of the hard work, time, and money you've invested lead up to that final moment of reward. However, the work does not stop after the chop. It goes way beyond that.

The next thing growers do is to ensure that their harvest is processed correctly to prevent any chances of it becoming damaged or rendered non-smokeable. Drying and curing cannabis flowers post-harvest is an essential measure to minimize the risk of mold contamination. We wouldn't like to get our Cannabis plant spoilt at this final stage. Not when its consumption is just a few days away.

Drying and curing will also greatly improve the taste of a crop. This is due in part to processes that break down chlorophyll over time, resulting in a less-harsh taste. This aspect is especially important for those aiming to share their product or use it medicinally. Drying and curing are also reported to reduce the anxiety associated with smoking cannabis; it may even increase cannabinoid potency.

Trimming Before Drying

I'm sure a lot of people would have it at the back of their mind whether to let their plants dry before trimming or trim it while they are still wet. Nevertheless, the drying process begins as soon as you cut down your plants and begin to trim the buds. In doing so, you will notice how sticky and wet the fresh flowers are.

Although this stickiness is a good indicator of the sheer amount of psychoactive resin on your buds, it also offers a great breeding ground for fungal and bacterial contaminants. Leaving buds lying around in this state is usually a recipe for disaster; so it is best to act with haste.

There are multiple ways to trim your plants at the start of the drying process. "Wet trimming" involves trimming as soon as plants are ripe. Cut off the branches one by one and proceed to use sharp scissors or shears to precisely trim down excess plant matter. Although the buds are of primary interest, the sugar leaves also contain lower cannabinoid levels and can be stored separately, then made, for instance, into edibles later on.

"Dry trimming" is a technique mostly used when a grower has a large amount of plant matter and little time to process it. This involves cutting off branches and hanging them whole from drying lines. Once the plants are dry, they are then trimmed and processed. Dry trimming is more challenging when it comes to neatness as small sugar leaves will have curled in toward the bud. Plus, dry trimming may cause a loss of resin due to the agitation of the branches when hanging and being handled.

Regardless of the trimming method used, it is important to process your harvest within a suitable drying room. A drying room should ideally feature a cool and dark environment between 15-22°C.

The Drying Process

If you opt to use the wet trimming method, have your sticky buds at the ready. Now, you will need to spread them out across a large surface area. Placing them directly on cardboard or newspaper is not advised, as this does not subject the flowers to total aeration. Placing them upon a dry rack with netting or wire mesh is a far superior option. This allows airflow to reach all sides of the buds. If possible, use drying racks large enough to spread the buds out evenly, with a few centimeters between each one.

Using a small rack means piling buds on top of each other, which may result in uneven drying and possible mold contamination. When buds are left to dry, they start to release a lot of water. If this water cannot escape, pockets of moisture may begin to form. Moisture is one of the multiple variables that mold needs in order to proliferate.

Be that as it may, this chapter concludes the end of our amazing cannabis experience. You've successfully started the journey of growing your plant; you've taken the liberty of looking after the plant, nurturing them, and tending to them; a little patience in drying and curing them after harvest won't be a bad idea.

Conclusion

Wow, I'm sure you will agree with me that it has been a beautiful experience as you browsed through this handbook with rapt attention. We are also aware that before you picked up this book, you had been thinking growing cannabis is a big deal. Well, we hope we have cleared that perception. On the contrary, growing cannabis is fun!

It is exciting as well as rewarding. Now, before you start growing momentum and getting eager to start your cannabis farm, please ensure you check what the laws on cannabis are in your region. We wouldn't want to enter into this venture blindfolded.

Sow your seeds, tend to the plants, and harvest them when they are due, it's that easy. Even as a beginner or expert grower, we opened up your eyes to the possibilities around you when it comes to growing cannabis. But first thing first, remember the golden rule – Never tell anyone you are growing cannabis.

No one can be trusted. Even your best friend can rat you out. Keep the secret to yourself. There is a joy that comes with smoking your own cannabis as that is the end product of your labor. The flowering stage, like we discussed, is the most sensitive part of the plant's life cycle. Endeavor to give your plant the most needed attention during this stage.

Chapter by chapter, we discussed how to make your cannabis plant grow healthy and bear more buds. Some people might not entertain the idea of using a chemical like fertilizers to boost their plant. That is understandable, and there are other ways in which you can boost the rapid growth of your plant without resulting in the use of chemicals. Explore those ways we mentioned.

Notwithstanding, the issue of security is very vital in this kind of business. We are sure you must be getting used to that word as we used it in more than five instances in this handbook. Security is the bedrock of growing cannabis. If there is no security, your farm is highly vulnerable to everyone out there. Imagine a situation where everyone can go in and out of your cannabis farm with little or no security check. Not cool, right?

Focus on the brighter side. There is no business that doesn't come with risk. As a beginner, you might try growing cannabis more than once but end up getting frustrated with too many pests and small predator problem. Even before your plants reach flowering stage, these pests may have killed it off, making your effort looks useless.

Well, I've got news for you. Do not feel dissuaded. This handbook will help you through these obstacles.

Finally, you have all you need to grow a healthy-looking cannabis in one shot. Follow the steps, procedures, and processes we've discussed in the chapters of this book and the sky will not only be your limit but your starting point.

Good luck and God bless!

Marijuana, Cannabis & Weed 101:

The Ultimate Guide To Marijuana Growing, Investing, Business, Stocks, Addiction & Horticulture - Including Cannabis Spirituality, Extracts, Medical Uses & Chronic Pain.

By Simon Major

Introduction

When we are making reference to marijuana, we shouldn't forget that growing marijuana is not just a hobby or killing time; it is, in fact, a booming business for others in this line of the venture. It goes beyond putting a seed beneath the ground. It goes beyond taking in a few clouds of smoke indoor with your friends. It is a means of survival for some.

The moment you realize this notion about the cannabis plant, the clearer your perception and narrative about the psychoactive herb would be. It is no doubt one of the most effective and multi-purpose plants grown in the universe. Its uses and benefits have been felt in both ancient and recent times.

Now, let me share the turning point of my life with you. Like everyone else, I was once an average stockbroker who had a thing or two for marijuana back in the 90s. Due to one or two reasons best known to me, I hadn't paid real attention to it. Like you, I had always felt marijuana was only meant to be smoked and forgotten.

I learned I was wrong when I came across Marcus, my very good friend. He had been a small-time cannabis indoor grower who was always looking over his head due to the criminal laws levied against the psychoactive herb. He explained the economics and numbers behind the cannabis plant.

He made sure he explained everything to me in detail; the stereotypes, the massive benefits, and the good side of the plant. As a small-time stockbroker, I knew I had hit jackpot. With my little knowledge of economics, I knew investing in this line of business would bring back good returns.

Coupled with the fact that many other states in the country had been removing the ban placed on marijuana, my urge to throw money at this venture increased - and in the end, produced more than enough returns, which changed my life for the better.

This book would serve as an eye-opener and a well-detailed guide toward enlightening you on the economic, medicinal, and environmental benefits of the cannabis plant. Not everyone knows the importance of this psychoactive herb to mankind. Nevertheless, I will walk you through every detail and step.

Mind you, the end result and goal of this book is not to force or cajole you into believing what you don't wish to believe about the plant. Instead, it is to prepare your mind and help you see the cannabis plant in a whole new dimension, thereby changing the narrative and stereotype you are stuck with.

In the end, you will come to realize and accept the fact that growing weed is one of the most lucrative and booming business one can invest in. Allow this book to take you into a whole new experience with the new face of medical marijuana. You will be shocked at how developed and advanced medicine has become, especially with marijuana being the centerpiece of that achievement.

With that, I welcome you to my world of marijuana. Relax, take a deep breath, and learn from my experience!

Chapter One

Historical Background of Marijuana

When weed was discovered in the ancient days, it was strictly used for medicinal purposes. It was highly cherished and completely reserved for nobles and the privileged. That is how important cannabis became during that era.

In the old Chinese Empire, the herb was used to treat mostly the privileged (emperors, nobles, etc.) as it was a very rare and extremely expensive plant during this era. Also in Ancient Egypt, the plant and its seeds were used in treating pharaohs and most members of the ruling council.

With these instances and various examples, you would agree with me that the cannabis plant was highly cherished as one of the most beneficial plants in the world. It's funny how a plant whose origin is known to be mysterious spreads its roots across the nations and kingdoms of the world.

Although many scholars and researchers agree with the fact that cannabis got its origin mysteriously with its seeds, leaves, roots, and even flowers spreading all over the universe with the help of dispersing agents. For example, when birds perch on the ground, they are likely to pick up a seed or two with their claws or beaks, then dropping it off in another region as they fly off.

Herds of cattle grazing off a particular area might also get their hoofs attached to few seeds as they walk on the plants unknowingly, thereby, planting the seeds as they run or walk off. These mean, among others, led to the discovery of the rampant state of the plant in the world. We wouldn't be entirely wrong to say that these diverse agents of dispersion played a large role in the discovery of the cannabis plant in the world today.

Ever since it was discovered, the plants have been studied and researched to determine how it can be further used in a more advanced, improved, and modernized form; tablets, syrups, and so much more. Marijuana was first used in liquid form when it was added to wine by an old Chinese doctor in the 1800s. It has taken many form since then.

Let's take a brief look at the history of cannabis in different ancient empires and kingdoms.

The Chinese Empire

Long before civilization came to the west, the old Chinese empires were known to be far ahead of other kingdoms and empires of the world. They had developed their own alphabets; they had learned to read and write; they advanced in terms of medicine and science, and so much more. In Asia, the Chinese empire was no doubt the only powerful hegemony with lots of subsidiary kingdoms and empires. That is how well they were developed.

Marijuana started out as a normal type of plant until the physicians of that era realized the plant could mean so much more if refined and reprocessed. Little by little, marijuana became the secret recipe of most of the medications prescribed in that period of time. One way or the other, its multipurpose usage got out, and it in turn became a very scarce herb, especially with the importance and high price tagged on it.

During this era, it was believed to have been used to cure severe illnesses like heart failure, skin disease, cancer, indigestion, and so many more. It became even more popular amongst the nobles and privileged of the society in that era. Due to its expense, it could only be afforded by the nobles. Physicians of the era also made improvements to it.

Some other uses of the plant during that era are the use of hemp for clothing materials. At a point, it became the secret weapon of the Chinese in winning wars and battles. Hemp was heavily used to make thicker and stronger fabric, which covered both the people and the soldiers.

For example, there has been archeological evidence which is believed to be around thousands of years old and contains details of how hemp was heavily used for clothes. The Shu King and old poetry also laid emphasis on how China made use of hemp in creating fabrics.

With the firmness and vigor of the plant, it was highly used by the Chinese in archery. Arrows, bows, and bow strings were made out of hemp. They are much stronger, better, and flexible, thus helping the arrows cover more ground when fired. China can be said to have categorically reached a highpoint in the use of cannabis.

The Greek & Roman Empire

It wouldn't be fair to discuss the historical background of the cannabis plant without delving into the contribution of the Greeks and Romans toward this psychoactive herb. According to ancient history, it was believed by some botanists that the plant was given recognition during the Greek-Roman era as ancient botanists like Theophrastus (287-371 B.C.) made reference to it in their ancient writings.

Though there had not been many writings on the medical benefits of the marijuana plant during this era, they had acknowledged the fact that cannabis is one hell of a plant with lots of outstanding properties. They had further laid emphasis on the seeds and leaves of the plant. The old Greek-Roman writers had made us understand that no part of the plant is entirely useless.

From the buds down to the leaves, from the branches to the stems - all the way to the roots, it's all useful. Additionally, it was popularly known that the man who coined the word "cannabis" is an old Greek historian known as Herodotus (490 A.D.) As a traveler who was in love with history, he had moved around the areas of old Egypt, Babylon, Tyre, Thrace, Scythia, Arabia, Palestine, and many others.

He had also written lots of works on the cannabis subject matter. In one of his works, he had described the cannabis plant and seeds as a vapor-emitting plant. Other notable Greek-Roman philosophers, historians, botanists, etc. had given their own view and opinion on the plant. With this, you would surely agree with me that cannabis seeds and plants have been in existence for over a long period of time.

The Old Egyptian Kingdoms

During the days of the pharaohs, Egypt can be said to have has been at the forefront with advanced technological, medical, social, and psychological feats which made other kingdoms around them bow to their might. The story of Joseph in the Bible should serve as evidence of these assumptions.

Medically, old Egypt found solace in the eyes of marijuana. Over time, they were able to get the optimal benefit of the plant in the best possible form. As a form of medicinal purpose, the cannabis plant, as found in the ancient Ebers Papyrus (a book penned down in the 1550 BC), was used to cure lots of illnesses, both chronic and mild.

The Ebers Papyrus placed emphasis on how to prepare the herbs, when to apply the herbs, and where and how to apply them. It shed more light on how marijuana can be used to move past depression and other psychological issues one might be passing through during this era.

As early as 2000 BC, this medicinal herb was used for curing diseases like glaucoma, cataracts, hemorrhoids, vaginal bleeding, and so many more. Even when the cannabis plant does not entirely cure these diseases completely, it goes a long way in ensuring that the diseases are reduced to the barest.

In Ancient Egypt, marijuana moved toward a whole new dimension entirely. It no longer served the purpose of just being a cure or psychoactive substance; it graduated into a way of life. It turned into a movement. It became a culture and a religion to many old Egyptians. For example, archeologists have found traces of weed in the tombs of notable personalities in Egyptian history.

This only signifies one thing; cannabis had indeed eaten deep and made a profound mark in their society. It had grown to be generally acceptable by everyone, including the pharaohs of the era. Now, picture cannabis as an aspect of their religion - makes sense, right? If their pharaohs could accept the plant and even be buried with it, isn't that an evidence to our religion narrative of the era?

Aside from that, the pictures of the goddesses of the old Egypt (Goddess of Wisdom and Goddess of War) Seshat and Bastet also have the cannabis plant in them. This is even stronger evidence toward categorizing the cannabis plant as part of the religion of that era. As a way and process of fulfilling the religious beliefs, festivals, and culture of the Egyptian era, cannabis was highly used. It is also believed to have been used for witchcraft and rituals.

The Islamic Kingdoms

It is important to know that the Koran and the Prophet himself didn't really lay a direct emphasis on the prohibition of cannabis. However, it is still a haram (forbidden) in the religion of Islam - just as pork is forbidden. This doesn't change the fact that the cannabis plant had been imbibed into the cultures of diverse Islamic nations and kingdoms with historical backgrounds up to a few centuries ago.

Many Islamic countries like Afghanistan, Morocco, Iran even turn this plant into hashish. In recent time, the possession of these plants in many Muslim countries like Saudi Arabia and the United Arab Emirates is met with the harshest of punishment. Thus, even if these Islamic kingdoms and empires acknowledge the existence of the cannabis plant, they have been able to ban it in their environment.

Nevertheless, some kingdoms still made good use of the cannabis plant, most especially in treating diseases and making hashish. Let's take a look at old Morocco, Turkey, and Iran, shall we?

Morocco, in the 7th down to the 15th century, had been characterized by a lot of tribal and cultural activities, before the introduction of Islam into their system. Cannabis growing really thrived under this era with lots of businessmen and women moving into Morocco with caravans.

The next century witnessed a change in the face of cannabis as the then king ordered the large production of the plants. Farmlands were given out, seeds were shared, and planting techniques were taught to people during that era. What followed after this was surplus cultivation of the plant all over Morocco, though only practiced by a tribe.

This was, however, curtailed by King Hassan I when he issued new laws and restrictions on the cultivation of cannabis in the 18th century. Instead of the art of growing cannabis being reserved exclusively for a particular tribe, it now became an art that could be practiced by more than four different tribes. This boosted the economy of Morocco.

The modern-day Turkey, which is now characterized by a high number of Muslims, has a long time history with cannabis. With the diverse tribes and religious beliefs scattered all over old Turkey, cannabis production had been acceptable during that era. There has been evidence such as paintings, old texts, writings, and medicinal formulas as far back as 100 BCE.

Even the Ottoman settling in Turkey made good use of the cannabis plant. According to ancient history, the 1631 Era of the Sultan Murad IV ushered in a severe punishment levied on the possession of tobacco, wine, and coffee. But interestingly, he allowed weed and opium to remain legal. You can see how much of importance the cannabis plant was to this era.

In old Iran, the connection between cannabis and its people ran quite deep. The Scythian tribe, which were quite popular in old Iran starting from 700 BCE, had been known to be quite synonymous to the psychoactive herb. Their name has appeared in more than a dozen ancient texts. They mostly used the cannabis plant for recreational, medicinal, and even spiritual purposes.

Just like the Scythians, other tribes and religions, like the Zoroastrians, also recognized and made use of the cannabis plant for religious and medicinal purposes. It was even indicated in their holy book, Zend Avesta as Bhanga, the Sanskrit meaning of marijuana. Afterward, the production of cannabis fell heavily under the introduction of Islam.

Rulers of that period now started imbibing the Islamic culture and principles in their way of life, ruling, and relating with their subjects. Many social events and activities were now seen as a sin. Alcohol was banned and taverns were closed. Gambling was prohibited and gambling dens were closed down. Prostitution was banned and whorehouses were shut down completely. And the use of cannabis was placed on direct persecution.

The origin and historical background of this plant is just as mysterious as its healing properties. This chapter should be able to open up the pods of questions in our minds. Among these questions should be, if these old empires with very shallow, crude, and backward settings had fully understood, unraveled, and made use of the benefits attached to this plant, why are we fighting against it? Why are we depriving ourselves of these same benefits? These and more are what the next chapter will explore. You don't want to miss it.

The historical background of the cannabis plant is so interesting, one would start thinking he or she had actually lived through the era. The great warriors and emperors of the era had actually found solace in the plant. This is how important the plant was to these kingdoms and empires. Now, what is the current state of the marijuana plant in the present world? There is only one way to find out.

Chapter Two

The Face of Marijuana Today

Without mincing any words, those who had fought vigorously against the legalization of the cannabis plant would have come to agree that the cannabis plant had been doing more good than evil so far. When we take a deep look at the giant strides and footprint cannabis plant had made for itself in the field of modern day medicine, we would surely come to a conclusion that the psychoactive herb is more a blessing than a curse.

The people of the ancient days were no fools, especially when many of them devoted their time, life, and attention toward unraveling the mysteries and charming properties of the plant. It is to be known that the cannabis plant has moved past the traditional phase and stereotypes in many regions of the world today. Many enlightened people no longer see the plant as something deadly and harmful, as it was painted to be.

One way or the other, the advocates of the plant stood through difficult and trying times. They knew and believed in the goodness locked up in the plants and continued to propagate their belief in the best possible way that they could; mass media, protests, rallies, internet, and so much more.

We can now say that lots of people are enjoying the dividends of this struggle. With over 20 states in the United States of America and many other countries legalizing the use, growth, distribution, and possession of cannabis, one can boldly say that the new face of marijuana is that of liberation and freedom.

Freedom and liberation from oppression, the right to engage in any peaceful activities concerning cannabis, and so much more is what many people enjoy in these regions. The road toward this liberation is a very long one. From the 1800s, to the 1900s, and into the 2000s, it has not been an easy venture in the United States.

According to the opinion of many scholars, when the use of cannabis broke out in the United States, it was widely and generally acceptable. At a point, it was seen as the most effective cure and alleviator to many diseases and infections. For example, marijuana was used as an alternative to pharmaceuticals in curing diseases in many hospitals. Doctors had begun to try out medical marijuana and the result was overwhelming.

Aside from medicine, the cannabis plant had also been mastered, especially when it was first used in the fabric industry in the early 1800s. These and more were the early uses of the cannabis plant. If we are to properly go into detail on the relationship between the United States of America and marijuana, then you should know that this book's pages might not be enough. There are lots of events, activities, and important occasions that characterized this relationship.

According to scholars of this field, this relationship dates as far back as the colonial era. There had been an encouragement and acceptance in the overall production of the plant in old America. The government had even approved and made sure hemp was produced in large quantities for it served as a very big part of the economy (basically used for ropes, sails, and fabrics).

The end of the Civil War saw a decline in the production of the cannabis plant. Hemp was now gradually losing its popularity as regards fabric production, while medical marijuana flourished even further. It was now been processed into tablets and syrups, sold freely in many pharmacies during this era.

During this period, recreational use of marijuana was at it barest with the population's focus on pure medical marijuana. However, the end of the Mexican Revolution of 1910 saw the rush of Mexicans into the United States with recreational marijuana introduced into American society.

This was the beginning of the "evil menace" stereotype tagged on the plant. Natives now believed that the marijuana plant is a clear sign of the Mexican vices which were brought with them while crossing the border. They believed the "newcomers" brought this menace that was eating through their social virtue.

Then and there, the crusade against marijuana gained life. The government and another capitalists whose gains lied with banning cannabis ensured they manipulated every text, churned up every intimidating idea, campaigned for cannabis outlaw, and promoted the negative narrative of the cannabis plant.

Every arm of the government was used to exert fear into the minds of the people. The mass media was not left out. Lots of articles and over-exaggerated reports were produced to scare the people. During this time, a popular movie "Reefer Madness" made it to the screen and it played a very big part in making the world see the cannabis plant as an evil and harmful herb one must cut off in order to live a meaningful life.

All this later gained the much need result by 1932. Over 20 states in the country had been able to successfully ban and criminalize the cannabis plant. One would have thought these advocates of criminalizing cannabis would stop there, but guess what? They went further. They championed the Marijuana Tax Act which was passed into law in 1937.

This law would see to the fact that marijuana use was prohibited and if it was to be used, then the tax placed on it would be quite enormous. All efforts to make the government see reason fell on deaf ears. For example, during the Second World War, the United States went back to hemp production as it was used for creating military gears, parachutes, and other war equipment.

Additionally, there had been lots of studies and research carried out to counter the assumptions and arguments of the advocates of this prohibition law. For instance, the New York Academy of Medicine went into research and came out with a detailed conclusion that the cannabis plant does not induce or lead to addiction or cause insanity or psychological stress.

But the minds of the capitalists and United States government was made up. There was no turning back as they put the last nail in the coffin. They knew many of their assumptions and arguments were wrong. Over and over again, research had been able to show that marijuana is not as bad as they had painted it to be.

They knew this, but it became too late to turn back. Thus, the 1950s saw another new set of federal laws which were all directed towards the criminalization of the plant. Sentences and persecution were now being carried out on anyone caught with marijuana. They now became offenders in the eyes of the law.

After a brief period of actual struggle and strongholds persecutions, the 1960s saw a whole new trend in the issues related to cannabis. The law became even more accommodating when former presidents of the United States, President Kennedy and President Johnson authorized and gave legitimacy to a report that showed that marijuana does not and cannot in any way induce addition or make an addict out of anyone.

Believe it or not, it gave an actual voice to the advocates of this plant. This became a reference point for them in their arguments. In the 1970s, many of the condemning laws and edicts were abolished and lifted. This loosened the noose around the neck of the advocates. It became a step they took toward achieving their set out goals and objectives.

For example, the Commission of bipartisan Shafer inaugurated by the former president of the United States, President Nixon came out with a conclusion that the decriminalization of the recreational use and possession of the cannabis plant should be put into law. Notwithstanding, Nixon strongly rejected the views - but more than ten states in the country followed this conclusion by lifting the ban on personal use of marijuana.

The struggle didn't stop there, though for a while, the tides were definitely in the favor of the cannabis advocates along 1972 down to 1975. People actually started seeing marijuana as something far from harmful. They started accepting marijuana and hating the stereotypes weaved around it.

However, this brief break-through was short-lived with the revolt and movement of the concerned American parents in 1976. They believed the legalization of the plant would really affect their kids mentally, physically, and even psychologically. The parents still believed that cannabis is very harmful.

To this effect, the War on Drugs crusade of the 1980s was inaugurated. This pushed back the good work lots of advocates of the plant had been doing over the years. President Reagan even initiated the "three strikes, you're out" program to help facilitate the laid down laws and also help exert fears in the mind of the people as regards cannabis.

This program meant that if you were caught once or twice with the possession of marijuana, you would still be given the light sentences of a few months or years. But, a third time would give you an automatic and non-negotiable life sentence. The War on Drugs crusade also dragged on to the President George Bush era in 1989.

Since then, the acceptance of the cannabis plant in society has been quite easy. There was a complete change in public opinions and polls. This was major as a result of the new face of marijuana was showcased. Instead of the harmful and addictive substance it had been perceived to be, marijuana gradually became the new face of modern day medicine.

For example, when the HIV/AIDS virus was freshly discovered in 1996, the medicine line saw a new savior in cannabis as it was processed and refined to tackle this virus. California, as a state, passed into law Proposition 215. This would legitimize the sale of the cannabis plant to patients of the virus, as well as its medicinal use.

Aside from this virus, the cannabis plant also became the major and most effective panacea or alleviator to many other deadly diseases out there. This is one break-through many can't deny. The line between the criminalization of the plant and legitimizing it became slimmer ever since. With many states putting the stereotypes and old merciless laws behind them, marijuana is now adding new feathers to its cap.

From its recreational use to its medicinal use, the plant has been effective and efficient in many ways. In the modern world of today, marijuana has enjoyed and is still enjoying a lot of freedom and privileges as compared to many decades ago. Aside from the normal reports and little charades put together by its antagonists, marijuana has so far been great.

Culturally, you would agree with me that marijuana is very strong in this aspect. Some tribes and regions of the world are now being familiarized with the plant. For example, Jamaicans and African Americans in the United States. Many believe that marijuana is a strong part of their cultural background and it has grown to be a very big part of their history.

Even Mexicans and some parts of the music world, such as rap, are believed to be moved and built on the psychoactive herb. Just like the old Egyptians, the cultural view of the cannabis plant in today's world is in sync with religious beliefs. Many still believe the plant to be linked with their religious beliefs and some see it as a total haram (forbidden).

Economically, the marijuana plant has given hope to the hopeless and even put food in the tables of many. According to statistics, many states in the United States of America that legalized the use of cannabis recorded a great amount of economic boom.

Colorado, for example, made a total of $135 million in tax on marijuana. With the sales and distribution going a little over $990 million, this has been a significant increase in the revenue of the state as of 2015. According to reports, there has been more than 8% increase in the local economy.

Job wise, marijuana has created lots of job opportunities over the years after many states started accepting it. For states that passed a bill on Medical marijuana, nurseries and dispensaries were formed which in turn led to the creation of many jobs and appointments.

A lot of added hands were employed to make sure medical marijuana became a success - from doctors to scientists, experts, analysts, and so much more. Marijuana became an industry on its own. The impact of this job creation has been felt on the economy ever since.

According to an RCG Economics and Marijuana Policy Group study in Nevada, there has been an increase in the number of created jobs since 2015. The report also shows that recreational marijuana has been the a backbone to as many jobs as possible in the country - over 40,000. Then, since the legalization of the cannabis plant in more than 20 states, there have been over 80,000 jobs created.

And according to more research that has been conducted by the New Frontier, job creation with cannabis is predicted to be on a whopping amount of over 1.1 million as of 2025. By this time, virtually all the states in the country might have legalized marijuana either for recreational or medical purposes, thus allowing it to quickly spread all over the nation. There would be more jobs, like farmers, processors, distributors, sellers of the cannabis finished products, and more.

Medically, it's only a blind man that would say the benefits of the cannabis plant aren't glaring. There have been many breakthroughs over the years in the line of medicine with marijuana. Although we will discuss more on this topic in the following chapters to come, here are a few key points we should take note of as regards medical marijuana:

1. Marijuana is the new medicine.
2. Marijuana breakthroughs in medicine have been a blessing.
3. It has served as an alleviator or panacea.

Marijuana didn't just start gaining acceptance from the general population all of a sudden; it was a gradual step and phase. Nevertheless, the future holds great promise for marijuana with lots of in-depth research and studies carried out to push forward the positive notion of the psychoactive herb. Do you think cannabis is addictive? Do you think it is a step toward the use of deadlier drugs? Do you still follow this bandwagon and stereotypes? Our next chapter will shed more light on this topic.

Are you surprised? Are you shocked by the fast surge of developments and improvements within this short time? That is the power of medical marijuana. Even at that, many people still hold the thoughts of the cannabis plant being addictive and this is what our next chapter will talk about.

Chapter Three

Unveiling the Truth Behind Cannabis Addiction

It's crazy how many people still hold the traditional view of the cannabis plant. No matter how hard you try to spin the web of grace and innocence around the cannabis plant for some people to see the benefits and good the plant contains, they would still see the plant as a harmful and addictive substance.

This stereotype had become a tough stain to wash away in society - the society now associating the plant with other vices. This is a total misrepresentation and misconception of the plant. The plant doesn't in any way instigate mediocrity or madness as many are led to believe.

The cannabis plant doesn't breed addiction in any way whatsoever. This is a misconception the advocates of the cannabis plant have spent lots of years arguing. There have been lots of research and studies that have shown this argument to be untrue. Addiction comes from within, and not from the plants.

Allow me to enlighten you on this by pointing out the conclusions of psychiatrists as regards cannabis addiction. In this line of thought, there are laid down rules and classifications of addiction in all psychoactive substances and herbs. You would be shocked when we start listing them all out.

Meanwhile, the notion of cannabis being addictive and making a menace out of society should be totally scrapped from our minds. This was basically used as a tool or strategy to cloud the public minds and get their support in the fight against the plant. Thanks to a series of campaigns and crusades, the cannabis plant got the crucifixion they had hoped for.

A very good example is the release of the movie titled "Reefer's Madness". This movie portrayed the wrong side of the cannabis plant and painted the plant black in the minds of people. To cut it short, it advocated that the cannabis plant would give nothing more than an addicted, delusional, and abnormal society.

Other campaigns like adverts, protests, and open talks have added nothing but fuel to the already-established fire. These people now begin to quote ancient philosophers, medical practitioners, and botanists of the cannabis plant out of context. They started twisting the words of these people to suit their campaign.

Like I pointed out above, the field of psychiatry has undergone a study which has been compiled in a single document named ICD-10 (Classifications of Mental and Behavioral Disorder). We will only be touching a few parts of the whole document that concerns this chapter. Thus, if you are interested in reading the whole document, then here is the website to it – http://www.who.int/whosis/ied10.

Lots of medical practitioners out there can do little or nothing at all because the government of most parts of the world has placed an indefinite ban on the use, distribution, and possession of the cannabis plant. Most times, even research and studies that ought to have been carried out in establishing and finding a solution to issues like cannabis addiction are either discouraged or even scrapped.

A lot of these governments, especially in the countries of the third world still don't believe that there can be a lot of good in the marijuana plant. Aside from the medical breakthroughs the plant has had in recent times, it can also be a booming power for their economies as it would bring money to the table as well as reduces unemployment.

For a medical doctor to be able to help a patient who is struggling with the cannabis herb, he or she must have had even the slightest of experience with cannabis-related problems. But these experiences can't be easily gotten when the government prosecutes such offenders and smokers of the plant.

Additionally, no one has written a clear book prescribing what to do with someone having cannabis issues. Scientists and pharmaceuticals have also not been able to develop a drug to match addiction in the lives of an addict.

Now, the question we should ask ourselves is, if there is an addiction that can be created from the properties of the cannabis plant.

The next best thing any sensible medical practitioner can do in this situation is to go ahead and gather data and information from other places, particularly other countries. For example, you might want to know how this kind of situation was dealt with in Japan, or even Nigeria. However, what you would see would only be based on the use of the psychoactive herb and its abuse.

This medical research and studies would say the same thing no matter how hard you dig. They all would answer the same questions with the same answers. Questions like, can the cannabis plant be abused? Can the cannabis plant be addictive? Can the cannabis plant addiction lead to loss of lives? Has cannabis intake generally led to death? What treatment is effective for treating cannabis addiction?

The answer to the first question is Yes. Quite a lot of people abuse even normal drugs, not to mention a psychoactive one. Be that as it may, this should not be on the cannabis plant. It should, in fact, be on the user of the plant. The answer to the second question is also a Yes. The cannabis plant can be quite addictive. But it's an addiction is not as a result of the properties of the plant but the unending desires of the user.

The answer to the third question is a No. The cannabis plant's addiction has never recorded a single death. This is to tell you how safe the cannabis plant can be. Additionally, the answer to the fourth question is No. The cannabis plant does not in any way contain a harmful chemical as it has been portrayed by some people.

As a matter of fact, only one person in the world can be said to have undergone a death by marijuana consumption. In the end, the cause if the death was later found out not to be of marijuana intake but of another particular type of herb he had taken before taking the cannabis plant. Reports showed that his body had an allergy to that herb and he didn't know about it. Lastly, the most effective treatment cannabis addicts can get is psychotherapy.

Key Reasons Why the Addictive Tag Still Holds

1.   It is difficult to tell if one is under the influence of cannabis, especially while doing risky jobs and performing dangerous activities like driving, weight lifting, and so much more. Most people take this psychoactive substance as a stimulant in garnering an insurmountable kind of prowess or creativity in their hobbies, jobs or activities.

While a musician can successfully and effortlessly smoke up a large part of a skunk without feeling down and get creative and inspired as a result, it is dangerous for a tank driver to stay on the road while smoking a large chunk of the cannabis plant. He or she might lose focus.

Believe it or not, there is not much difference between a person that smokes cannabis and ends up getting the wheel afterward and an alcoholic driving a car after getting wasted. Both are highly irresponsible and not wise. To this effect, only addicts can do something like this.

What makes it even worse is that there is no direct breathalyzer that would show how much cannabis one has taken or even show us who had cannabis under the last minute. No such technology has been invented so far. Thus, it becomes difficult to trace and tell if one is smoking or smoked before taking the wheel.

2.   Generally, it is believed that most pot smokers are definitely lazy. According to research, it takes an extra effort or pushes from smokers of the cannabis plant, most likely addicts, to stand up and perform their daily routine after sitting there to smoke. Nature is orderly and regular.

Thus, what comes after smoking the cannabis plant is a particularly sound sleep. This is synonymous to most smokers out there. There would be an increasing push toward being lazy and lackadaisical. Instead of standing up to get productive, they often become heavy sleepers and dormant people.

Cannabis smokers are also seen as dependents in society, especially addicts. The point of this addiction comes with being docile. That way, all they would now start thinking about is to smoke, eat, and watch television. Nevertheless, in cases where the addicts don't have a permanent way of depending on and leeching off of others, he or she would definitely have to put up with not smoking for some days, so as to earn something.

It is important to note that curbing the effects of the cannabis plant takes just a few days of not smoking at all. Unlike other stimulants and psychoactive substances that hardly wear off and make one a chronic addict, the cannabis plant's effect would eventually wear off with one looking brand-new, as if he had never touched a leaf before.

But if a cannabis addict can't seem to get his or her life back after a few days of staying clean, then we would recommend you visit a psychiatrist or a psychologist. In no time, you will be back, just like new.

3. Currently, it is very glaring that the government had misled the general public into holding negative thoughts of the innocent plant. It's unfortunate how a super-power would be allowed to be used inappropriately by some people, just because they don't see the goodness and benefits the plant holds for mankind.

But guess what? The government later found out that they were wrong about the plant. Medical marijuana is a start. There had been lots of breakthroughs and inventions from the cannabis plant as it takes modern day medicine to a whole new level. It has given a cure or serves as an alleviator to a lot of deadly diseases out there. For example, cancer, Hiv/Aids, and many more.

Yet the government still covers their mistakes up with the word "addiction." Inasmuch as they know the cannabis plant is far from being an addictive substance, they still hold on to the stereotype as their only justification for leading the public astray. Additionally, if cannabis was legalized and its criminalization removed, then a lot of people jailed under this offense would have to be released. The laws would have to be changed and this would affect a lot of departments which would, in turn, lead to loss of jobs.

This are the key reasons why the cannabis plant must continue to look bad so that the government can continue to come out clean looking like the saint and impartial judge it has always been.

Cannabis addiction at best can be said to be at the hands of the user. The addiction and dependence of the user on the drug is entirely not the drug's fault. However, this idea has been misconstrued and twisted to suit the narrative of the people against the cannabis plant. I'm sure this chapter had been able to clear your mind of this misconception. Don't you wish to know more about this psychoactive herb? The next chapter will focus on marijuana horticulture. You don't want to miss it.

Chapter Four

Marijuana Horticulture

Horticulture as a field of study has proven to be more challenging and intriguing of late. With lots of trending topics and important issues rising up in this field, horticulture has risen up from being one of the most underrated fields to an exciting one. And marijuana is one of the few reasons why horticulture is gaining momentum in the modern world.

The growing of crops or plants can be said to have been in existence as long as man's existence as a whole. Right from the beginning, growing of crops has been one of the productive activities men have engaged in so as to keep their circle intact. Gradually, what started as just a means of survival has now grown to be what has been keeping the world alive for centuries.

Who would have thought this same cultivation of crops would turn out to be essential in the life of man today? Research is now being carried out as to how to further improve cultivation. Horticulture cuts across diverse crops and plants, some of which are very peculiar to the benefit of mankind. For instance, the cannabis plant.

The cannabis plant is one of the most multifunctional and multidimensional plants you would find in the world today. As we have shown in our previous chapters, this plant has had a long history of struggles between its advocacy and with those that hold a negative view of the plant. Aside from the rigorous tiffs which had been going on for a long period of time, the plants are now making waves in modern society with lots of outstanding and spontaneous discoveries.

Marijuana horticulture focuses solely on the agricultural basis of the cannabis plant. How do we grow the cannabis plant? What exactly do we need to do to breed a healthy plant? How do we maintain growing a good plant? These and so much more are what horticulture as a field of study is going to delve into.

A lot of cannabis experts and writers have written lots of books on this subject matter. Little wonder why everyone can start up a marijuana farm or indoor garden as easy as can be. The books are there to help you through. There are also a chunk of online materials and articles you can rely on; and you can also get tips from your experienced friends.

Be that as it may, we are going to be discussing what marijuana horticulture entails. Without leaving any stone unturned, we are going to walk you through the basics of marijuana horticulture, so that when you start growing your marijuana, it will be as easy as a walk in the park.

The Grow Room

The grow room only applies to the indoor growing of the cannabis plant. The grow room is a cool and dry space in the building that is purposely used for growing the cannabis plant. As a beginner, this is one of the first things you should consider before starting your own cannabis garden. Ask yourself these questions: How secure is your grow room? How far is it from the living room? Does it contain the smell?

These and many more questions are what you should ask yourself before choosing a strategic room for this purpose. When growing in the room, be sure it is spacious enough for the plants to thrive together. And need I remind you that the lighting needs to be topnotch? A grow room with a good setting would definitely breed nice, healthy plants.

Now, how do you setup your grow room? Setting up the grow room and knowing how to arrange the grow pots in perfect rows can be quite tedious. You don't think one can just drop the grow pots anywhere in the grow room, do you? You need to be exact and strategic in this setting.

The size of the grow room solely depends on the number of cannabis plants you would like to cultivate. Also, it depends exactly on what you really want to use it for. Recreational purpose of the plant mostly leads to small-time gardens. But a medical purpose of the plant is certainly a big thing. Sometimes, it involves hectares of land.

The plants need to be arranged in a perfect manner so that your plants won't start competing for the basic and necessary conditions to grow. A bad grow room setting ends up in an unhealthy, uneven growth of the plants. They start getting intertwined, they start fighting for light, air, and water, they also compete for fertilizers and nutrients.

Like we pointed out above, the grow room is only needed when you are planning on growing your cannabis indoors. Everything needs to be intact. From the lights, down to the fans that would provide your plants with the air they need, the natural and artificial fertilizers, and so much more.

The grow room should also be very well-ventilated. Many breeders and growers would prefer their grow room to have a nice, big window. This comes in handy in case they might want their plants to enjoy light directly from the sun and also absorb natural air from the wind. Be that as it may, a big window also comes with big responsibility.

You should always be aware of how tall your plants can grow. Always make sure they don't become too tall or grow too close to your window. If you planning on keeping the security of your little garden topnotch, then always watch out for this. Or else, people or prying eyes from the outside might just be keen on what kind of business you are into.

Additionally, setting up your own grow room is no joke. It comes with quite a few financial decisions. With the least amount of $100, you are definitely good to go. So, ask yourself if you are really that buoyant financially. Aside from the money involved, there is a lot of labor too; cleaning, maintenance, and repairs. Are you up for it?

Hydroponics

Some growers of the plants prefer this method of growing the cannabis plant amongst many other methods. Hydroponics defies the traditional use of planting the cannabis plant. Instead of making use of everyday soil (clay, humus, sand, loam, and others), hydroponics focuses on using a kind of water solution that is filled with nutrients.

The use of hydroponics was first introduced when the growers of the cannabis plant began to find fault in the soil used in growing it. Instead of the plant growing healthily irrespective of the techniques used, it grows weak and the yields were quite below expectations.

Thus, it became pertinent for these growers to develop and initiate another method which would bring them out of the mess, cap their effort with success, and put a smile in their faces as they harvest – Hydroponics was born. Hydroponics would give the grower a lot of time to focus on other things as it doesn't require much labor. But it must be properly maintained from time to time. Hydroponics are mostly used indoors and here are its benefits over the old traditional soil method of planting:

1. While hydroponics can be recycled, the soil can never be recycled. If the soil gets contaminated or laced, then you would have to automatically change it.

2. Soils can be easily contaminated with diseases. They are also prone to pests problem, unlike the hydroponics method.

3. With the soil, one can find it very difficult to properly ascertain the right combination of nutrients needed. It is quite difficult to know if the Nitrogen content is not enough or even too much. This is very possible with hydroponics.

4. The quality of the soil determines what your harvest would look like. A poor soil definitely would lead to a poor harvest. Thus, one would need to constantly keep improving the soil, thereby involving more labor. Hydroponics doesn't take one through such stress and additional labor.

Hydroponics are easy to start. It doesn't matter whether you have experience in it or not. It is quite easy to execute and you would be amazed at the size of your harvest in the long run. Don't get it twisted, it also has its own crossroads and complex ways, but in the long run, you will figure it out.

Now, the question is how does one set up the hydroponic method of growing cannabis? It's pretty simple. There are lots of hydroponics methods which can be used or are essentially used by lots of growers out there.

Curing

Imagine curing your cannabis plant before harvest? Sounds crazy, yeah? It's just impossible. You need to first cut down your plants, clip the leaves off the stems and branches, and dry them in a separate room, before thinking of curing them. Curing the plant is a whole new technique that allows you to enjoy the best flavor, scent, THC, and so much more.

How do you begin curing your plants? I'll tell you. Before going down to this stage, be certain that your plants are already dried, most especially the buds. Be sure to protect the buds from pests or unhealthy conditions during the drying stage. That might affect the properties of the buds and end up giving you bad weed.

It is important to know that speeding up the drying of your plant is very bad for every part of it. It's bad for the buds, it's bad for the leaves, and it's bad for the trim. It would end up giving the plant a very bad taste. That would be like losing all the essential properties of the plants.

So, it is advisable to hang your plants upside down, especially with the buds. Dry them in a room devoid of light, air, and water. That way, the plants would dry up easy and fast. Gradually, the leaves will start drying up as well as the buds. When the outer parts of the buds dry up completely and the stems become bendable, you are good to go.

Like we pointed out above, the slow drying procedure is the right way to go. It will help your plants become the best in their properties and qualities. Patience is all you need. Trust me, I know it all. We tend to get impatient, especially when our plants are at their harvest stage. All we now start thinking about is how and when to reap the fruits of our labor.

Speeding up the drying process would cause your harvested plants a lot more harm than good. This is where curing comes in. Curing might end up helping you to improve the quality of the plants, especially the buds.

Hash-Making

Are you new to the term "hash"? Then allow me to explain it to you. Hash-making, which is also known as hashish, is the end product from marijuana after carefully separating the main plants away from the trichomes. These trichomes are mostly found in the flower part of the cannabis plant.

Therefore, you need to take very good care of the trim, so as to get enough trichomes for hash. Hash-making started out in ancient Morocco with lots of tribes making a living out of this venture. It is important to know that these trichomes can also be used to make more edibles aside from hashish.

In no time, you will also be a great maker of hashish. Marijuana horticulture will give you the necessary information you need to know about the different terms and concepts associated with the cannabis plant. That way, you would be able to have a concise idea before delving into it. The next chapter will explain the methods of growing the cannabis plant in detail.

Chapter Five

Indoor and Outdoor Growing

Far from the popular terms that surround cannabis growing like hydroponics, cloning, hash making, the grow room, and so many more, becoming a grower or breeder of this plant is quite rewarding and can be very exciting. Instead of buying from a local dispensary or organic farm, how about taking up the challenge of growing your own cannabis plant?

Growing cannabis can be a hobby for some people, while to some, it might be a means of survival. Irrespective of how and why you are venturing into this line of business, one thing still remains clear – there is an insurmountable kind of thrill and excitement that comes with growing cannabis, especially for the first time.

It's like procreation. Giving life to a single or group of seeds by just forgetting where you had buried it. However, many growers of the plant believe and hold the thought that growing the cannabis plant takes more than just forgetting to dig up the buried seeds. It takes passion. It takes love and attention. It takes care and support. It takes maintenance and repair.

Additionally, one needs to be prepared both mentally and financially before embarking on this journey, especially as it will involve a direct and personal involvement in every phase or stage. There are also important questions that need to be answered before starting your cannabis farm.

You should ask yourself if you have what it takes to pull this off. Ask yourself if all appropriate measures have been taken already. These and many more are the key areas and questions one need to really ask oneself. And if the answers are affirmative, then there is nothing stopping you from becoming the newest grower in town.

Be that as it may, there are basically two types of growing methods one can apply while growing the cannabis plant; the indoor cannabis growing method and the outdoor cannabis growing method. The choice is entirely yours to opt for any one of the two as both are very effective and highly efficient.

Where the indoor cannabis growing method focuses on growing the cannabis plant indoors, particularly in a grow room, the outdoor method specializes in growing the cannabis plant outdoors amidst natural events and circumstances.

However, they both have their own pros and cons. Where one can mostly grow the Indica strain in the indoor growing method because if its small size and height, growing the Sativa strain would be quite a disaster because of their enormous size and height. They would really want a lot of space that the indoor growing method can't provide.

The Seeds

First of all, before any growth can be carried out, one needs to get a good-looking and quality seed to sow. Like we had explained in the previous chapters, there are three types of species in the cannabis plant. These are the Sativa, the Indica, and the Ruderalis. These three are all great and vary in their properties and characteristics.

The Sativa, for example, are a strong, enormous, and gigantic species. The Indica are average in height and contain more THC content that the remaining two. And lastly, the Ruderalis are very normal with their contents and their leaves are more green than the others.

But if these three species are all great and perfect, how then do we select one seed species among the other two? This is entirely our decision. We can make our decision by looking at which seeds would fit and suit our plans and environment better. We can also make our decision based on the kind of growing method we will follow or the financial capabilities of our pockets.

Thus, how do we select the best of the seeds? It's pretty simple! With over 450 different kinds of cannabis seeds out there, we might find it a little bit hard as a beginner to select the best out of the rest. However, there are certain qualities one can use in knowing which seeds will be best when planted in our garden.

To begin with, we should look at the quality of the seed banks. Where did the seed come from? Can the breeder be trusted? If he can, then there is nothing stopping you from going for that kind of weed. Secondly, we should endeavor to check the flowering times of the seeds. It the flowering times go in line with your expectations, then there should be nothing stopping you from buying that seed.

Thirdly, you can tell if a seed is good and would be fertile by just looking at the physical outlook. If the seed looks wrinkled and quite old, then such seed is not a good seed. But if a seed looks fresh and still exudes its color radiantly, then that is the kind of seed you should probably start using in your farm.

Factors to be Met Before Growing

There is a golden rule among all cannabis growers which must be adhered to no matter the circumstances – Never ever tell anyone that you are growing the plant in your building or in your backyard. If you want to stay clear from the relevant authorities and prying eyes, then we would suggest you take this seriously.

Aside from this, there are certain factors that must be met before starting out this hobby or business, as the case implies. As we all know, the federal laws and laws of over 15 states in the United States still have not come to terms with the benefit and good work the plant has been doing in recent times. Thus, staying stealth in those regions is the best way to pull off a cannabis farm.

However, staying stealth doesn't come easy and cheap. Its success largely depends on the important procedures and steps that had already been laid down in shielding the farm from necessary pests. This is where meeting some important factors before commencing your farm comes in. What if the relevant authorities check in after an unexpected surge in utilities? They might know what you are working on that is eating up so much electricity. What would you do if that happened?

Or what if a passerby reports your farm that is always on lockdown to the police and they come snooping around. What would you do? These type of questions are exactly what we need to answer before starting our garden or farm.

Thus, security should be our number-one priority when choosing a place or building to use for this purpose. We should always make sure our farm is highly secured, so as not to put us in trouble or even worry about getting caught. And if it's an indoor growing thing, how secure is the grow room? The grow room should be situated in a strategic position where not just anyone can be aware of it or even get access to it.

Secondly, the environmental factors also need to be checked. What is the environment like? Is it weed-friendly? Or the same old stereotyped environment? That way you will be able to know how to conduct and run your farm without even alerting the slightest threat.

Thirdly, we should also lay emphasis on the climatic condition of the areas or environment we intend to use. We all know too well that most plants end up not growing in severe weather conditions like excessive rain, drought, and other harsh weather conditions you might not think of. To this effect, always have it in mind that you need a place with great climatic conditions to grow your cannabis plant, especially if you plan on doing it the outdoor way.

Lastly, we need to be financially buoyant before we can even start thinking of pulling such thing as growing the plant off, to begin with. Money does it all, no matter how severe the problem seems to be, throw money at it and watch it diminish. You will need money for lots of things if you are the indoor growing type. Thus, how prepared are you financially?

With that being said, this chapter will enlighten you on the methods of growing cannabis with these two trusted methods. It will put you through the task of becoming a beginner in the area of growing cannabis with any of these methods conveniently and effortlessly. Never forget that the choice is yours to choose which will be best for you.

Indoor Cannabis Growing

The indoor method of growing the cannabis plant is quite popular amongst small-time cannabis growers and those that grow the plant secretly due to the iron hands of the law in their region. Nevertheless, it is an effective way of staying stealth as a cannabis grower. The indoor growing method requires quite a lot of money before one can successfully embark on it.

With lots of things to buy; for example, the lighting, artificial fans, artificial soil, nutrients, chemicals, and so much more. Lighting is the most important thing when it comes to this kind of growing. We would recommend you choose the lighting source very carefully. You can either go for the artificial lighting that comes with bulbs and fluorescents or the natural lighting that comes with direct sunlight from the windows.

We should have it in mind that natural light may not be as productive as artificial light. While the artificial light helps the plant produce big thick buds and leaves, natural light hardly produces big leaves. Also, while natural light is free of charge, artificial light will cost some money.

Artificial fans should also be placed in strategic areas of the grow room. That is the only way your plants can take in air. Sowing the seeds in the indoor growing method is just like any other type of sowing method. Get your grow pot, dig the soil to a preferred length, before burying the seeds in the soil.

Afterward, you wait for it to germinate while caring for the sowed seeds. After a couple of days, the real challenge would start as your seeds must have sprouted up the surface of the earth.

Outdoor Cannabis Growing

Growing the cannabis plant outdoors is like growing any other plant you can think of. But this time, with more than enough security. It is only advisable to grow the cannabis plant outdoors in regions and areas where growing the plant is legal. Also, if you have the license to grow such plants, then there is nothing stopping you from growing cannabis outdoors.

Aside from these reasons, one needs to be attentive and be vigilant when growing the cannabis plant outside. People might tend to get the wrong idea about you or your little farm. Some might even go ahead and call the cops on you or even weed out your cannabis plant in your absence. Obviously, they would think it's all about saving you from trouble.

Some people might just end up stealing your plants if there is no security in your outdoor farm. Occasionally, you will have a problem of teenagers ripping off your plants behind your back. This may be as a result of their addiction to smoking weed or for financial gain. If this kind of thing happens to you, what would you do?

Calling the cops is out of the option. How about dealing with it the normal way? Tighten security. You can also disguise your cannabis plants amongst other plants. That way, people would find it very hard to identify which is weed and which is the regular plant. This is one of the safest and cheapest means of securing your plants from thieves and the relevant authorities.

You can also build a kind of brick fence which will serve as a barricade between your plants and passersby. This should hide the crop well from prying eyes, thus, keeping it safe from outsiders. Nevertheless, outdoor growing entails sowing your seeds outside with natural conditions and settings influencing its growth.

The kind of species that are mostly grown in this method of cannabis growing is the Sativa species. The Sativa species are giant, tall, and full of leaves from the top to the bottom. They mostly prefer open space to grow out perfectly without restrictions. Unlike the indoor cannabis growing method that is confined to a grow room, the outdoor promises unlimited space to the plants.

There are lots of techniques that can be used in this method of growing cannabis. For example, thinning, cloning, trapping, and so much more. The outdoor growing method of the cannabis plant welcomes as many plants as you can grow. It all depends on your capabilities and ability to provide security for the planted seeds. You can even grow as much as a thousand hectares if the need arises. The choice is entirely yours.

Be that as it may, the indoor and outdoor methods of growing the cannabis plant are particularly useful in the verge of creating your own farm or garden. Irrespective of whichever method you are opting for, all you need to do is just to follow the steps and guidelines as explained in this book, or even get tips from experienced growers.

Never forget that growing the plant doesn't come easy at all. It involves lots of stress, care, attention, and maintenance. You need to be sure that the water is not too heavy for indoor growing. You need to make sure the plants grow in uniform height, so as some would not block off sunlight that is meant for all.

You need to pay attention to the pests (both human and natural), as they will keep coming, even when you least expect them. Get all the vital nutrients for each stage, so as to boost the plants and keep them healthy as they move from one stage to another. These and many more are the things that would need to be done as a beginner.

And trust me, these won't come easy as a novice. Sometimes, you might even forget to feed the indoor garden nutrients or change the soil when the need arises. You may also forget to give water to the outdoor plants or even help spread pollen on the female plants. Do not panic in case your plants start showing signs of abnormalities after being forgetful of some things. It's not your fault that you don't remember. Instead of beating yourself up, why not make the necessary adjustments? It's never too late to correct one's mistake with the cannabis plant.

In no time, you are going to move from the novice that you are to an expert in growing the cannabis plant. Instead of consulting the Internet or even books on growing cannabis, you will be an expert grower of your own. It's that easy. Being consistent, focused, and knowing what to do at the right time is all it takes.

Chapter Six

Sowing Phase, Vegetative Growth Phase, and Flowering Phase

It's only natural for one to know what comes next after having an idea of how to go about growing the seeds. In the previous chapter, we laid emphasis on the two methods of growing cannabis; indoor and outdoor growing methods. We also noted that it is entirely your choice to choose which would be more preferable. But if I were you, I would go for the one that will leave little impact on my pockets.

Like we pointed out in our previous chapter, the indoor growing style is great but would definitely incur more cost than the outdoor growing method. The reason for this deduction is absolutely obvious to even to the blind. While the indoor growing technique comes with costs for things like the lighting, the artificial fans, the artificial soils, the cabinet, and so much more, the outdoor growing method comes with free-of-charge natural conditions like the wind, the sunlight, the fertile soil, and so much more.

Thus, be sure to ask yourself which will be very suitable for your condition so as not to inconvenience you in the course of growing your weed. That way, you would be able to come up with solutions and repairs where needed in the garden, either financially or even mentally. That is the only way you would have a clear mind in carrying out the necessary steps for a successful and healthy plant.

It is important to be conversant with the phases or stages of the cannabis plant. Just like human beings grow from child to teenager, then to adult, so does the cannabis plant also take form as it sails through different stages while growing. Each stage is delicate and should be approached with care and attention.

Be that as it may, there are basically six stages in the life cycle of the cannabis plant, but we will only touch on the most important three. Not that the others are not important or even vital to the growth of the plant, but as a beginner, I don't want to bore you with too many details. I'm sure, with time, you will also fill in the blanks yourself when you grow from a beginner to an expert grower.

Remember Marcus, the guy who introduced me to this money making venture? He gave me an important tip when growing the cannabis plant and this had helped me sail through the stages effortlessly, even when I was just a beginner like you. He had asked me if I was married. Just the way you would be dumbfounded at this question is how I was too when he dropped it.

How was being married relevant to the growth of the marijuana plant? Then went further by explaining what he truly meant. He said if I am married, then I should make the cannabis plant my wife's rival. That is the only way I can care for it. And if I am still single, then congratulations, I just found myself a sweet little green bride.

In other words, the marijuana plant should be taken like your new bride. It should be paid attention to and cared for immensely. Just as you don't joke with your wife or partner, we shouldn't joke with the cannabis plant. This is the plant that can either elevate us on the social ladder or criminalize us in society. Now, the choice is totally yours on how you choose to take care of your own cannabis garden, either indoor or outdoor.

So what are the three important stages we would most likely see in our plant? Can they be altered or tweaked? Can they be hastened or slowed down? What if we want to skip one or two stages; is it possible? As a beginner, your mind will probably be running through all those questions.

Possibly, a feeling of impatience will engulf our mind as we tend to really anticipate every stage of the cannabis life cycle with excitement and much enthusiasm. For example, we would most likely be eager to see our seeds germinate and sprout seedlings when we first sow them as a beginner. Trust me, I know the feeling because I have been there before. Even in the slightest delay, we tend to get worried with lots of questions running through our minds.

Did we really sow the seeds properly as instructed? Did we dig too deep? Was the water too much? Or the sunlight not bright enough? When this happens, you really need to take a chill. Relax your mind and don't get too worked up. The seeds will definitely germinate, even though it might take time. If you read the breeder's instructions well, then you would see where it was written that the seeds might really not take shape as the breeder has earlier suggested.

Thus, just believe in yourself and every other thing will fall into place. From the sowing stage to the vegetative stage, and lastly flowering stage, it's all connected. It's like a synergy that builds up, leading from one phase to another. Want to know more about what to do when these stages arrive and how to know each stage? Then let me enlighten you.

The Sowing Phase

This is the first step to grow every plant out there. All plants or trees come to start with their seeds, no matter how tall, big, or even giant they may become afterward. The same thing goes for the marijuana plant. All the species - the Sativa, Indica, and the Ruderalis are grown from their seeds.

So, therefore, when we have gotten the right quality seeds and strain we would like to grow in our farm, the next thing to do is to sow them. There are lots of ways to sow cannabis seeds. By it's entirely your choice to choose which one would be more preferable to you. You can follow the traditional way of sowing it in the soil. This is mostly preferred by lots of old fashioned cannabis growers out there. It's easy and precise.

Other methods of sowing the cannabis plant include the seed towel method, where the seeds are folded up in a moist clean or cloth for germination. The Rockwool method is also very effective and efficient. These are new techniques tried by adventurous growers and they are quite great for propagation and transplanting.

In this phase, the seeds open up in the soil after few days, spreading out its fresh roots in the soil. That way, it would make sure it holds the soil as a balance before pushing up the seedlings to the surface of the ground. When the sun continues to shine on the premature leaves, water been fed to it, and the nutrient of the soil giving it strength, it would have enough tenacity to withdraw from the seed totally. It will push the shell of the seed away.

This is when the seed has fully germinated into a seedling. This is also the part where you start jumping up and down after seeing the two premature leaves sprouting off the ground. Trust me, no feeling beats this kind of living. It's the evidence that proves that you are doing the right thing so far.

From here, you have nothing to fear. That is, if you care for your seedlings well. They are what will gradually develop into a full plant with time. The sowing stage doesn't require much stress. As a matter of fact, they don't even need much water, nutrients, or even air. Keep everything moderate. If you are operating an indoor garden, then you should keep the soil moist, but not wet. That is exactly what the seedlings need to thrive.

Mind you, some seeds would definitely germinate before others. You cannot stop nature from happening. But when this happens, you can only allow them to thrive. After they had reached a certain size and height, you can trim them to be uniform. That way, none of them will take all the light at the expense of others.

Be that as it may, your seedling will start growing gradually until its roots are fully firm in the soil. That is the first step toward vegetative growth. Only a firm root can hold up a full cannabis plant like Sativa, for example. Afterward, the seedling starts forming a shape with the stems getting thicker and the branches spreading out. The leaves would also start growing out more and more.

Though the stems would definitely be weak and fragile, we would still advise you to use a support system to hold on to the stems and maybe the branches too. You can use sticks or any other thing to hold them. If you are growing your cannabis plant outdoors, then you should consider using something stronger to hold the plants because of the strong wind.

After some time, the lower leaves would start dropping off with new ones formed. The leaves would obviously be getting bigger and greener. The stems would also be getting thicker and stronger. When all this happens, welcome to the Vegetative Growth phase.

The Vegetative Growth Phase

This stage is obviously the intermediate stage in the life cycle of the cannabis plant. Just like the teenage stage is the intermediate stage in the life of a human, so is the vegetative growth stage in the life cycle of the cannabis plant. In this stage, the plant is fully ready to spread out its full length, height, and width. All it needs is more than enough food power to realize this.

Trust me, the plant is very delicate in this stage. The amount of light, air, and nutrient you give to it in this stage will largely determine how much bigger or taller the plant will be. So if were you, I would focus on giving the cannabis plant the best treatment I could offer. That is the only way your plants can be healthy and full of the right content.

The vegetative growth stage would definitely fill the plant up with the right energy and cannabinoids. All the plant would seek to do is to keep taking in the food, light, and water thrown at it. I would encourage you to be very careful also in this stage. Always be vigilant as to be sure the plants aren't taking in unwanted nutrients or even harboring pests that may be harmful to the plant.

In the vegetative growth stage, the plant will grow continuously till it reaches a preferred height. For Sativa, they may be up to 12 feet tall. The stems will also get thicker and stronger with time. The leaves will definitely get fuller and bigger. Some will drop off or dry off while giving room for others to grow. The plant will also start showing you signs of its sex.

The vegetative growth stage can be altered, but it all depends on which growing method you are adopting. For indoor growers, they can cut off some parts of the plant or even tie up some parts which seem to be bending towards the light. This can be said to be a plant growing out of their respective space.

Additionally, this stage in the life cycle of the cannabis plant makes it highly vulnerable to outer bodies like pests and diseases. You really need to pay very close attention to your plant during this stage. Since they are looking great and all green, pests like molds and mites will definitely want to live off the plant. It is your job to always make sure your plants stay clear from pests.

Additionally, when your plants start growing tall in the grow room or the outdoor garden, they might want to grow over the window of your grow room or even grow over the fence on your farm. It is your duty to make sure this doesn't happen. They might put you in danger when people see the plants, even from afar. You might not know who would report you to the relevant authorities or even rip off your sweat from the root.

Thus, if the plants start showing signs of full-blown maturity - if it doesn't grow any taller than it already is, and if it leaves or stems are thick, wide and strong enough, then we are definitely approaching the last phase of the life cycle – the flowering phase.

The Flowering Phase

Before we would go fully into this phase in the life cycle of the cannabis plant, it is important I take you through the development of the plant toward the preparation of this phase. When the cannabis plant moves pass the vegetative growth phase, it only signals the next. That way, its features and body qualities would tend to change and take a new form. Some parts of it would start modifying toward new development.

The fast-growing stage of vegetative growth slows down gradually. Instead of the plants growing tall and thick like it had been for weeks now, it would now start to focus on another part of the plant entirely. It will focus its energy in other aspects, like creating more branches with nodes growing all over them.

This is one of the steps the plant will take in preparing for the next phase. Remember when I emphasized the importance of feeding your plants well during the vegetative growth phase? Well, I wasn't joking. The size of your plants rest on how well you feed your plants.

Additionally, the plants will fill out more and it will most likely take the shape of a Christmas tree. This is when the branches will bring out something called Calyx. Calyx will only appear at strategic positions and places; most likely, where the branches and stems meet one another. When this happens, then I would suggest you start preparing for the flowering phase of the plant.

The flowering phase is where everything comes into light. It is where you would be able to differentiate the sex of the plant. While the male will produce smallish grape-like balls that are joined together, the female will only produce pistils at the top of the plant. This is one of the easiest ways to differentiate which is male and which is female.

I don't need to tell you that the plants will fill out even more with flowers decorating every corner? Mind you, this phase, even if beautiful to behold with the booming flowers hanging at every corner, can be quite displeasing to the nose. The stench of the smell can come across very strong. It might even penetrate the grow room or the outdoor farm, thereby putting you into trouble. Thus, finding a strategic place before growing becomes imminent.

Additionally, the male plant will spill its pollen all over the female during this stage. The pollen can also be gathered and stored if you want to. You can help spread the pollen on the female plants if the male plants didn't get to spread the pollen properly, after the flowering stage comes to the harvest.

If your plants have successfully sailed through these stages without any qualms or even little qualms as it is your first time, then congratulations. It's really not easy pulling this off. It takes commitment and lots of patience to pass sail through this.

These phases are all important in the life cycle of the plant. Always be at attention in case where the plant starts showing signs of abnormalities. Now, if the plant passes through these phases, harvesting it will become the next best thing to do. How would you go about this? Our next chapter will shed more light on this.

Chapter Seven

Harvest of the Cannabis Plant

If your cannabis plant makes it to this stage, then congratulations. And if it doesn't, it's not the end of the world. Don't let it discourage you from achieving your set out goals. Harvesting cannabis is one of the most amazing feelings one can ever experience when embarking on a venture. After the rigorous stress that involves growing, tending, and caring for your cannabis plant, the harvest is the reward for that tenacity and commitment.

Take this scenario, for instance. Everyone wants to reap the fruit of their labor. This applies to even cannabis growers. You don't expect a grower who had toiled on his cannabis farm for months to end up losing out on the rewards. That is very painful to a businessman, to a grower, and to any sane person.

Thus, taking every means necessary, using every measure you can find, and doing everything in your power to make sure your cannabis plants turn out well is paramount. As discussed in the earlier chapters, pests will come knocking on the door heavily. Each phase of the plant might come with a whole new species of pests.

There are pests that will attack your plants from the top and those that will eat up your plant gradually from the bottom. Endeavor to always check out the cannabis plant from time to time. Treat it like a new bride. After all, it is a business venture which you've invested your money, time, and attention into.

Additionally, don't get ahead of yourself. Many beginners in growing cannabis end up jumping even before they leap. They end up counting their chickens before they hatch them. This is wrong. They often start counting and calculating how much harvest they have and some even go a long way to doing the math as well as churning out numbers.

I would strongly advise against this attitude as a beginner. Remember, this is your first time growing or attempting to grow the psychoactive herb. What if you end up making mistakes? And mistakes will definitely come as a beginner. For example, pouring too much water in the soil, getting the wrong nutrients and supplements, transplanting an infected plant unknowingly in the midst of other healthy ones, and so much more.

These mistakes would definitely spoil your plans, thus making you become hopeful for nothing. This will hurt your feelings and ego. It might deter you from attempting to grow the plant next time. Be that as it may, we would advise you to keep your mind open to the best and worst possible case, instead.

Now the question in your mind right now should be, how do you know when it's the right time to harvest your plant? If that is how you feel right now, then you are definitely in the right place for answers. Every stage and phase of the cannabis plant follows an appropriate timing. Things just don't happen for no reason. For example, you don't expect the vegetative stage to come after the flowering stage, do you? Just like nature is orderly and regular, so is the cannabis plant.

Notwithstanding, the timing must be perfect in order for you to get the best out of every cannabis plant in your garden. Too early might be quite disadvantageous and too late might just not give you the best quality you had hoped for. Here is what to know about the early and late harvest of the cannabis plant.

Early Harvest

Harvesting early might be great, but if the buds are more paramount to you than any other parts of the cannabis plant, then we would recommend you to take a chill before harvesting. Don't rush it, don't harvest early, and take your time so as to allow the plants to produce and come up with enough bud.

Nevertheless, there are lots of good reasons why harvesting early might be just what you should actually do with you cannabis plant. Bugs and pests might serve as a very good factor towards this. If your plants keep getting attacked by a number of bugs and pests or are showing signs of them, it's better to harvest early when the plants are still in good condition.

Outdoor growers, for example, are quite familiar with this type of harvest. The plant cannot be moved since it's already in the ground and not some grow pot. Thus, the harsh weather conditions, the different kinds of buds, and the human and natural pests that would come should definitely make your harvest early.

Though the THC content might be quite low as compared to the normal period of harvesting, it is still worth it. It's better to own maturing plants than to lose them all. Some growers also prefer to harvest early when their plant starts growing stealthily. Some also harvest early if the surrounding plants used as a cover of the outdoor cannabis start overpowering the cannabis plant.

When these plants start taking the necessary nutrients needed by the marijuana plant, it is allowed to do a premature harvest. Avoid it as much as possible, but if the circumstances revolve around using this method, then I'm afraid it leaves you with no choice than to start harvesting early.

Late Harvest

While lots of THC cannabis growers don't really subscribe to this method of harvest, some still prefer it. They believe purposely leaving the plants to stay alive even past the harvest stage gives their plants the best quality they hope for. This gives the plant a special kind of high which many smokers tend to love in weed.

With late harvests, one needs to be extremely careful so as they don't self-pollinate all over again. But rest assured, late harvest also comes with its own advantages. Be that as it may, it is totally your choice to choose from both methods. As we have explained above, both methods are effective and efficient, as it all depends on how you want your weed to be after harvest.

So, do you want the best psychedelic effect? Do you want the best level of THC? Do you want your cannabis plant to exude a pleasant scent and taste?

Then all you need to do is to know the appropriate time to harvest. Have you ever thought of balancing the properties of the plant to achieve the best result? That might also work like magic and give you what you need.

Nevertheless, if you ever find yourself in a position where you seem lost and confused on the timing of the harvest period, why not take a good look at the plant and search for key attributes the plant would exude. Here are some of the features you should look out for in the cannabis plant when planning on harvesting:

1. **Color changes in trichomes:** If you're looking for the best way to know if your plant is ready to through the sickle, then look no further. This sign is one of the most perfect and easy ones you will find when your plant is ready. Make sure to examine and carefully check the color of your trichomes.

The examinations cannot just be carried out with an ordinary glance but with the use of a magnifying glass because these trichomes are minute and extremely small. They are also known as resin-bearing glands because they help secrete resin which is vital during the flowering stage.

It is important to know that as tiny as trichomes can be, they also change state with time. At first, they can be very clear and solid. They start getting cloudy with time, before they develop into an amber state. Be that as it may, we would strongly recommend that you start your harvest immediately when you start noticing a mix in the state of your trichomes.

In other words, if your cannabis plant trichomes have a mix of different states, then you are good to go. If the trichomes contain a mix of clear, cloudy, and amber at the same time, then this is the perfect time to harvest your plants. In case you are wondering why can't you just wait for all the trichomes to turn amber, here is why. If all trichomes are amber, it means the plants would have a considerably low THC and a high risk of CBN, which is not cool for most growers.

2. **Leaves turn yellow:** When the leaves start turning yellow, aside from being caused by pests, unhealthy growth, and so on, it then signals one thing – harvest. It is, in fact, a great sign most growers of the plant expect

to see at the appropriate time. For other plants, if the plants start turning yellow, then its trouble. But for marijuana, if the leaves start turning yellow, then it's jubilation. Imagine the irony!

Watch out for the fan leaves. They hold the key to this method of knowing if the time is right for harvest. The fan plant will start turning yellow or even end up breaking off by itself. This indicates the plants getting weak and approaching the harvest stage. Some growers and breeders have made one or two complaints about this method of checking the harvest timing not being effective in their garden.

That is definitely because too much fertilizer was used in their indoor or outdoor gardens and farms. To this effect, we would advise you to examine your plants carefully and look for other available signs that might serve as an indicator.

3. **Pistils change color:** Although we can't guarantee you a 100% effectiveness and efficiency in this way of checking the harvest timing, after a careful perusal, you will notice that part of the pistils are brown or turning brown. This is an indication that the plants are ready for harvest.

Though this might not sound or look too great when you experiment it, it has been quite wonderful and extremely effective with many breeders. If the trichomes method doesn't go well with your plants, then try this one. Your plants might just be ready without you knowing.

4. **The leaves start curling:** Do you know that the cannabis plant gradually reduces the amount of water it takes in the more it approaches its peak (harvest period)? The moment this starts, the plant will start drying off and the leaves automatically start curling, too.

This is a very good sign in knowing that your cannabis plant is ready for harvest. Don't get it twisted, there are also different reasons why your plants can start drying and curling up. For example, pests can serve as a pain in the ass most times. Poor maintenance of the plants can also make this happen.

Be that as it may, if your plants start drying out and curling up even after you are sure it is very healthy, then you should know that this is a sign that your plants are due for harvest.

5. **Always check the breeder's timing:** In most cases, the breeders do make it very easy for you in their manuals or instructions. They make sure they stipulate every necessary detail on their seeds from the germination period down to the harvest. These might even contain dates and time frames.

It might be days or weeks, depending on the qualities of the seed you are planting. Thus, instead of just going through the extra effort and stress to search for signs and signals from the plants, why not just go with the breeder's manual? Nevertheless, these manuals and instructions might come out wrong, also.

For example, a breeder might stipulate a period of two weeks for the harvest period and the plants might be saying otherwise due to some factors; factors like the lighting in the case of indoor cannabis growing, or factors like the air, water and nutrients in the outdoor cannabis growing.

Be that as it may, the real question now should be, how do we go about harvesting our cannabis plant? Like many other crops, the cannabis plant should also be harvested by cutting it at the bottom (not too close to the roots). Both the indoor and the outdoor growing methods have their own similarities and differences.

These are definitely because of the processes and procedures involved in growing them, from the weather, down to the temperature, the water, climate conditions, and so much more. For example, you can go out in the middle of heavy rain and start harvesting your crops. Same way you can harvest in your grow room while it is still dark.

Harvesting the Indica Cannabis Plant

First and foremost, you need to decide when you want to start harvesting your plants. This should be the first natural thing to do before embarking on any kind of mission. You first make up your mind on what and when to execute it. That way, the rest would just follow naturally.

It is important to know that these different types of species of the cannabis plant vary in terms of their harvest. The Indica, for example, are known to be way smaller and shorter than the Sativa species, thus, require less strenuous effort when it comes to harvesting them. Picked a date? Then congratulations! Let's get on with it, shall we?

Harvesting your Indica plants, which are mostly grown indoors, is quite simple and less stressful. Make sure you use very sharp equipment or farm tools like the sickle in cutting the plant neatly from the bottom. Afterward, gradually pick them up so that the leaves might not litter your home and hang them up in a room which you've reserved for this purpose.

Make sure the room is cool, dry, and dark with no penetration of light. That is the only way you can keep the THC from degrading. The next thing to do after this is to take a clipper and start clipping the fan leaves, secondary leaves, trim or buds, and any other part that you feel will be useful to you.

Be sure to drop them in a separate pile so as not to mix them up. Each of them contains THC contents that are higher and more potent when compared with one another. After harvesting, you can now start curing the leaves.

Harvesting the Sativa Cannabis Plant

Just as we had pointed out above, harvesting the Sativa species is different from when harvesting the Indica plant. Sativas are generally large, tall, and giant plants with almost 12 feet of height and possessing up to 20 oz of buds per plant. Isn't the Sativa cannabis plant just amazing?

I'm sure you must be wondering if this species of the cannabis plant, which is mostly grown outdoors, is stressful and involves much work when it comes to harvest. Well, let me be frank. Yes, it does. As a matter of fact, it takes long hours and extra hands for you to finish up on time. However, it all depends on the size of your farm and how much time you have at hand.

So how do we harvest the Sativa species? Do we need to climb a ladder to clip off the leaves? Well, you don't need to do any of that. To begin with, you need to first get some clean sheets to place on the ground so as to make sure the leaves don't fall on the ground. Just like the Indica plants, the Sativa species should also be cut off from the bottom and be hung in a cool, dry, and dark room.

It is important to know that the Sativa plants are quite giant, large, and full of leaves. So, therefore, they should be cut into smaller bits before they are eventually hung. This would make it easier for you to handle them. Afterward, clipping the leaves comes next. Clip off the leaves and buds. You can also clip off any part of the plant that may be quite useful to you. These should be in different piles so as not to mix them up. Curing the plants should be the next thing to do.

Let me be frank; harvesting the cannabis plant is not an easy task, especially if it's the Sativa plant. They are very huge, giant, and tall. However, this would be automatically countered with the joy that comes with getting to smoke your own sweat. In case you want to start selling your harvested plants, the next chapter would tell you what to do.

Chapter Eight

Marijuana Growing as a Business Venture

In the previous chapters, we had expanded on what cannabis really entails and also how to grow the plant. As a beginner, you need not be told that that first time will always be the hardest. Even as you gather knowledge on how to go about it from books, the Internet or tips from experts, it won't change the fact that it's still your first time and mistakes are bound to happen.

Nevertheless, don't let these mistakes change your mind from what you love doing. Don't let it get in the way of your passion and hobby. According to statistics, at least 60% of growers of the cannabis plant cultivate the plant for recreational purpose. They see it as a hobby and mostly grow in the form of gardens.

Be that as it may, the new face of marijuana has changed this perception. A lot of businessmen and women now invest and venture in the growing of the cannabis plant. Just like I was ignorant of the financial path that had lied in this plant all along, many others also share the same view, especially with the stereotype revolving around the plant.

In early 2000, when the cannabis plant was gaining momentum due to its use in modern day medicine, the government of the United States of America began to reduce the bans and penalties placed on the plant. Businessmen like me who knew nothing other than how to make money saw a pathway that would only lead to one destination – more wealth.

Unlike the normal stocks and shares, investing in weed production and distribution is one of the most stable businesses you can ever find out there. Besides, the competition is quite few. With lots of people still engaged in traditional recreational activities, it would be a smooth sail for you toward reaching the top.

In recent times, more people have realized how much of a jackpot this business venture could be. This explains the recent surge in the marijuana business ventures out there today. From the stores, down to the mobile shops, medical marijuana farms, and so much more. This is a new trend.

Instead of just smoking the little weed you've grown, why not start thinking of establishing a business out of it? The marijuana industry is a newbie in the economy of the United States today. As a matter of fact, we can boldly say that the development and introduction of this industry are a result of the legalization and momentum weed started gaining due its medical uses.

As of now, over 25 states have legalized and removed any strings attached to the use of the cannabis plant in the United States. Thus, the decision made by these diverse states will definitely open doors of business opportunities for people in this line of trade or business. Buyers, sellers, and growers no longer need to the scared of doing what they know how to do best. Imagine what this would bring to the table of the United States economy.

When the industry was first introduced, there were lots of mixed feelings in the public opinion polls. When some were throwing their weight behind the arrangements, others were totally against it, sticking to their guns with values and morality. Until the financial report of the cannabis industry was published in 2016. The industry made over $6 billion which is totally mind-blowing. Only a few industries make as much as this figure per annum.

With that being said, I'm sure you would agree with me that the cannabis business is a very lucrative one. With lots of people finding a way or two to enter into this business, the business had been very helpful to the government as it provides jobs for people. The cannabis business can be seen as a recreational activity.

It gives you much convenience and time to do other planned stuff. It doesn't take your time and have you stuck behind a desk for hours like the white collar jobs out there. These are part of the benefits this line of business would offer.

Don't get it twisted, when I start using cannabis in some part of this book and marijuana in another, it doesn't mean I'm talking about something else. Both cannabis and marijuana refer to the same plant or weed. However, the original and scientific name of the plant is cannabis.

As a novice in this line of business, how about you start knowing the different strains and species of the cannabis plant? That way you will be able to decide which one you want to start banking on. The cannabis plant has three core strains, and they are; The Indica, The Sativa, and The Ruderalis. Aside from smoking the leaves and buds, the cannabis plant can also be used for lots of other things.

For example, they can be used for medicine, taken as hemp, or used as hemp in fabric making. Oil can be made out of it. These are a few things out of many that can be made out of the cannabis plant. Little by little, the marijuana industry is gaining momentum, especially with the stereotypes revolving around the plant being corrected and shunned.

This had opened up a pathway for many entrepreneurs out there to start considering this line of business without the fear of incurring a loss over their investment. There are now lots of entrepreneurs in the cannabis industry today. These entrepreneurs are capitalizing on every ounce of the idea they can get on the cannabis plant. They are making sure they fully exploit the use of the plants.

Don't be left out! This is an opportunity for you to make something for yourself. An opportunity to make your mark in the industry. Like I pointed out above, there was nothing meaningful I could have been doing if this business didn't change my life for the better.

Don't worry if you are starting small. Drop after drop, an ocean is formed. Set out your goals in the business and start planning on how to achieve them. Trust me, in no time, you will start seeing the dividends. If you are in the United States, within the 28 cannabis-legalized states, then get on with it, will you? What kind of business should you start within this industry? Here is our list of marijuana business ventures to engage yourself in.

Cannabis Business Ventures You Can Embark On

1. **Begin cannabis farming:** One of the amazing ways to get started is to grow your own cannabis plant. Instead of growing in just a pot or having a little indoor garden as usual, how about expanding? As far as we can tell, such laws that had been able to push lots of growers into hiding and limiting the number of plants they grow no longer exist in over half of the states in the country.

That way, you would be able to expand your customer base. Therefore, we would advise you to start fully as soon as you can. That is the only way to stay ahead of the competition. And if you are thinking of where to get the needed information, knowledge, and tips to start growing, then read through our previous chapters. We have already explained how to grow a healthy cannabis plant.

Our previous chapters would also show you all you need to know about the maintenance of the plants, how to really care for them in cases of pests and infections, how to grow the cannabis plant indoors and outdoors, how to successfully sail through the stages without qualms, and how to harvest them when the need arises. Always bear in mind that starting a weed farm openly would need a license unless you want to grow them in stealth. We would recommend you get one before starting anything.

2. **Processing the cannabis plant:** You don't want to soil your hands in chemicals, fertilizer, soil, and the stench of the cannabis plant? Then there is a suitable option for you, try cannabis processing. Unlike growing you cannabis plant yourself, what you would be dealing with is getting the raw plants from growers before processing them into finished products yourself.

All you need to do is to know how to really dry, cure and package the plants afterward before distributing it to its consumers. You don't have to be a big shot before starting out as a marijuana processor. You can get a grower or a couple of growers whom you can buy the plants from. Once you have established this arrangement with the grower, then the rest is in your hands. However, you would also need a license to become a legal marijuana processor.

3. **Marijuana deliveries:** Ever heard of transporting and delivering cannabis as a means of survival? Well, now you have. How about starting this business? Instead of going through the stress of growing the cannabis plant or processing it, why not be the man behind the wheels?

We all know growing marijuana can be quite stressful and productive. However, there are certain things we need to be sure of so as not to reduce the taste, effectiveness, and potency of our plants. This includes the environment and time. So, why not serve as a helping hand to the local growers as you become the only connection between them and their buyers.

What do you need to do? It's pretty simple. First of all, do your homework. Find the names of the local farms that buy cannabis strains as well as those that would be in need of the plants. Your job is to transport these plants. Many would agree with me that this delivery business of the cannabis plant does not really need much to start with. For example, the plants are lightweight, thus, there really is no need for getting a van. Even a motorcycle or bicycle will do just fine.

Additionally, your business should be registered. That way, you would not be involved in illegal dealings. Make sure you have a steady customer and client base. That is the only way to stay in this line of business. Feel free to make use of any form of advertising.

4. **Selling marijuana snacks:** In case you have not heard about it, marijuana snacks do exist. Most times, when the name cannabis pops up, the one thing that gets to our head is the smoking part of it. There are lots of things we can use the cannabis plants for aside from smoking them as weed. Making edible snacks from the plant is also one of them.

With smoking discouraged in the American population, people have now started getting creative with the plant. This is what gave birth to marijuana snacks. Marijuana can be added as an ingredient to any kind of dish or snack you might find out there. You can start small. You can start selling marijuana tea, cakes and pies laced with cannabis, and so much more.

5. Production of cannabis body products: As a multidimensional plant with diverse uses, the cannabis plant has proved to be highly beneficial to man. It has become part of the solution rather than the problem earlier envisaged. Scientists had unraveled the skincare properties of the plants long ago, thus, why not make use of this opportunity?

Start your own body care company where you can be producing your own cannabis skin care products like lotions, soaps, creams, relaxers, moisturizers, and so much more. Starting afresh in this line of business would be quite tedious and stressful, especially with the paperwork and procedures.

Thus, it would be much easier for you if you are already into this business. That way, you would only need to tweak the ingredients by adding marijuana. This line of business has been quite popular in recent times with lots of people delving into it. For example, Whoopi Goldberg, who is a popular actress has started her own line of skin care products with marijuana.

6. Open a cannabis accessories store: A lot of people are still living under the stereotype of cannabis being a harmful substance and it's users core addicts. To this effect, they find it extremely difficult to relate with users of the plant or even involve themselves directly so as not to be tagged an addict.

Well, there is still hope for you in this industry. Have you ever heard of a marijuana accessories seller? That is what you can be. You can delve into this line of business and sell different kinds of accessories to aid the use of the plant. These accessories, like the pipes, bongs, t-shirts, neck tags, dog collars, and so much more would help aid the consumption of the cannabis plant or even help propagate cannabis by its consumers.

You can also get creative with this line of business. You can create your own arts and craft, indicating or propagating cannabis. And if you are not a crafty person, you can purchase these things and end up reselling them at your own convenient price. All you need to do is to repackage and rebrand them properly.

7. **Plan marijuana parties:** This can be quite crazy, but you can be a marijuana party planner. Don't get me wrong, I'm obviously not referring to a party where underage and adults both drown themselves in the stench and potency of the cannabis plant. You would be surprised when you start learning that marijuana is actually being used as a theme in modern-day celebrations. For example, weddings, reunions, birthdays, and so much more.

Be an event planner and help your clients plan an awesome marijuana party amidst other celebrations. Additionally, you can also organize these kind of parties on your own and place a fee which should serve as the admission fee. Be sure to have a license for this in case the relevant authorities come calling.

8. **Start a cannabis flower shop:** You should know that some strains of the cannabis plant produce a kind of rare but beautiful flower at the end of the day. Imagine becoming a florist and selling these rare flowers for a price. Quite awesome, yeah? The little stench they come with can be taken care of but the bottom line remains that this kind of flower is very rare. Thus, why not utilize this opportunity to make money for yourself?

The flowers can be used for gifts during a celebration like weddings, engagements, proposals, birthdays, and so much more. If flowers are your specialty and what you are really good at, then we would recommend you check this business out. And if you have a floral business already, you should start thinking about including the cannabis flower into your types of flowers. With the stereotypes gradually diminishing and lots of people welcoming the idea of the cannabis plant with opened arms, there would be no better time to start this business.

9. **Become a marketer of marijuana-based products:** How effective are your advertising and marketing skills? If they are excellent, then this is the right business or you. All you need to do is to sweet talk your way into the minds of the public, making them dip their hands into their wallets or purses.

Be that as it may, the companies which are involved in turning cannabis properties into diverse finished products are always in need of people to spread the awareness of the good things revolving around the use of the plant. This can be an employment opportunity for you.

Are you media savvy? Can you whirl your ways around media marketing? Then start using that talent to make money for yourself. Make an arrangement with these companies. Discuss the terms and conditions as well as your remuneration. Consider this money for being amazing.

I am sure you didn't expect the cannabis line of business to be this sophisticated and enticing. I bet you are already calculating numbers and deciding on venturing into this line of business. Believe me, I was also like you. When I heard of the huge numbers involved, I started daydreaming. However, how do you become successful in this business? There is only one way to find out.

Chapter Nine

Steps to take in Starting marijuana Business as a Novice

The beginning of every great business venture we have out there started as an idea. Some might even go as far as saying it started as a dream. Whichever way you want to say it, both signal the same meaning. Ideas or dreams can come as a flash during a sober reflection of one's life as regards survival.

How can I survive this hardship? What can I do to stay afloat through this recession? These are the little questions that will keep popping into our head each time we face a setback toward making a good life for ourselves. And after we have tried lots of possible ways without achieving a breakthrough, giving up might be the only solution left.

If you are looking for a place to lay your head, make you feel valued, and tell you what your sins are and how you can overcome them, why not visit the nearest church? However, you picking up this book and not wallowing in your failures or licking your wounds behind closed doors points to one thing and one thing only; you want to make money. So let's get to it, shall we?

There is an African saying that goes like this; "If one doesn't get a peacock, then make use of an ostrich in its stead." In other words, if plan A doesn't work, why not start thinking of another plan? There is nothing wrong with tilting away from your set out goals. For example, if you get cut out of your job due to the resizing of the employee strength and size in your place of work, there is nothing bad in turning that disappointment into a blessing.

And when I talk blessings, I'm referring to a business. Why not consider starting a business of your own? You have the will since you're now jobless, the capital from your savings, and obviously, the time. All you need now is to think of which lucrative business you can start throwing your savings at.

Still haven't found one? Then we will recommend you start thinking of the cannabis industry. We would recommend you start thinking of engaging yourself in this money-making industry. With lots of money to be made in this line of business, all you need right now is knowledge and understanding. Experience would come with time.

Don't get it twisted, there are lots of business activities you can engage yourself in without getting to see a leaf of the marijuana plant. It's a pity a lot of people still hold the long chain of stereotypes and crucifixion of the plant. They believe that anyone associated with the word "cannabis" is no doubt a rogue, a smoker, and an addict. Thus, such persons should be avoided so as not to taint them with dirt.

Thus, if you are thinking along this line of thought, then there is absolutely nothing to worry about. In our last chapter, we listed more than enough types of business and employment one can secure or engage in as regards the cannabis industry, some of which doesn't even require direct dealing with the raw plants. These are cannabis marketers, distributors, and more.

Now, are you getting interested already? Have you started considering your future as a cannabis businessman or woman? Seeing your future tied to this line of business venture already? Well, very few people in the world today won't want to be, especially with the new rise in the numbers of the industry due to its legalization. It would take a dumb person not to consider this line of business and utilize every opportunity to make good, fast, and legit money.

More than 25 states and still counting have removed the bans and criminal laws placed on the cannabis plant today. In one form or the other, they have been able to see the light concerning this psychoactive herb. Instead of treating the cannabis plant like a plague, why not make money off it? This has been one of their ideas.

While some states only removed the ban on the cannabis plant as regards medical marijuana, others had gone ahead to extend this immunity over recreational, distribution, and lawful possession of the plant. This is, no doubt, a step toward the ultimate goal of every cannabis advocate – complete freedom for using the psychoactive herb.

Need I remind you that the year 2016 in American history witnessed a rapid surge in the economics of the cannabis industry? There was a whopping 30 percent increase in the overall returns of the cannabis industry with $6 billion. And according to forecasters, they have predicted that these massive returns will rise to over $20 billion in the year 2021.

Only those with foresight and great business minds would realize the opportunity this business line brings forth. Imagine how much this industry will bring to the table after passing these early stages since inception. In many years to come, only a few industries will be able to compete with the cannabis industry. Be that as it may, wouldn't you rather stamp your feet and write your name in the Hall of Fame of this growing industry?

Nevertheless, the road to becoming a successful cannabis businessman or woman is no easy road. If you think making enormous profit in this line of business and staying at the top of your game entails little effort on your part, then you are in for a surprise. It is important to know that there may be a series of annoying checks and balances, guiding laws and principles, taxes to be paid, and so much more.

These and many more would serve as obstacles and roadblocks to you in the course of your cannabis business venture. This is why we would advise you to take up a mentor in this line of business. Devote yourself to the knowledge of starting up and maintaining a cannabis business. There are lots of sources where you will find interesting facts and information that would aid you as you plan on growing.

You can take up books, for example, surf the Internet for articles and write-ups, and learn from the personal experiences of cannabis businessmen and women. This is the only way to hold on to your cannabis business. Trust me, you will need all the help you can get.

Therefore, there are steps to follow before starting a cannabis business. You can't just wake up one morning and start trading cannabis. This business and industry is a sensitive one, with a lot of stringent and strict processes and protocols that must be followed to the letter. The government of the United States has barely lifted the ban on this herb.

Thus, the last thing people in this industry would want to do is to ruffle up the emotions of the government.

To this effect, the procedures and ways of the industry are particularly strict and under hard time rules. Aside all this, it is as profitable as selling designer clothes. Be that as it may, there are lots of important and necessary steps one must take before starting up a cannabis business venture. But before we delve into this, let's take a good look at the necessary factors one must consider before starting up or even considering the cannabis business.

Factors to Consider Before Starting a Cannabis Business (Canna-Business)

Before starting out anything whatsoever, it is always wise to check the pros and cons. It is equally wise and important to check the kind of setbacks and obstacles one might incur in the cause of starting this line of business. This business might be even more complicated and complex than we had imagined. We tend to see this possibility if we don't plan fully ahead of time. If we forecast the factors that may hinder the startup process of our business, we would be able to tackle it with an alternate solution.

1. **Financial Strength:** Quite a lot of our dreams and ideas end up staying in our head due to this factor. Before conceiving the idea of starting your own cannabis business, have you tried calculating it costs? Aside from that, there are lots of things to do with money. Are you well equipped? You will need money for lots of things before starting.

You would need a van for transporting the products; you would need money to hire and pay staffs when necessary, the marketing and advertising costs, money for rent and utilities, money for supplies, money for your license, and so much more. Make sure you are well capable of this business before starting it.

2. **Residency:** According to the laws of cannabis in a country and as stipulated by the cannabis industry, one must possess at least 2 years of residency before applying for a license. This is a factor you should seriously consider. Are you qualified to be a resident? If yes, then it's all good. But if you are not, then that is a big problem.

Thus, we would recommend you get a license first, before even pouring your money into this kind of business. Nevertheless, if your residency is not up to 2 years, you can throw your weight around the industry. Maybe you might get lucky and your case would be treated with leniency. You never can tell.

3. Risky: Obviously, as an entrepreneur, this is the word that rings out every time you want to try out something new. What if you end up losing all your money? Especially being the first time you are starting this kind of business.

This line of business is extremely risky. It involves lots of unstable laws and regulations which might be in your favor today and totally be out of your favor tomorrow. That is how this industry works. However, choosing this path even after knowing all this is totally up to you. Even when entrepreneurs know that a turbulent storm might await them in the cause if this business, they still go for it as this risk comes with enormous profits.

4. Timing: If the timing is wrong, everything and every step you take to maintain and keep your business afloat would definitely prove futile. In starting any business, the timing matters a lot. Aside from that, the timing can also be delayed due to the paperwork and license.

Often times, it takes more than just a few weeks to get a license. For example, in California, it takes nothing less than 3 months and above to get your license processed. It all depends on where you intend starting the cannabis business. Now the question is, how patient can you really be? Take this in mind as you plan your cannabis business.

5. Getting Investors and Partners: Most times, raising the capital alone to start up a business can be quite difficult and mind disturbing. Little wonder why many entrepreneurs go to extra lengths to get an investment in their business.

Looking for investors in this line of business can be quite difficult. Like we said above, most people still see the cannabis plant in a traditional way. They always believed that engaging in cannabis activities alone doesn't really entail smoking it. Even throwing your money at it involves you with cannabis. While with some, their religion forbids them. That is why getting investors in this line of business can be quite difficult.

Steps to Follow Before Starting up a Cannabis Business

Like we pointed out above, every business whatsoever needs a laid down plan or steps to be followed before startup. The same thing goes for the cannabis business. You need to combine the right level of intelligence, labor, and capital in order to pull it off. If you can tackle the incoming obstacles that come as part of the laws and regulations of this line of business, then there is nothing stopping you from starting a cannabis business of your own. These are the steps to follow:

1.   **Write a Business Plan:** This is no doubt the first thing to do when starting out in the cannabis business. Getting a plan is the first step to every business. These laid out plans are the map toward success in any line of business. The plan should cover the expenses and running costs, the goals and objectives of the business, the milestones and timelines, and other such vital information.

The business plan would help you stay focused on the main purpose and objective of the business. If your business plan is brief, catchy, and straight to the point, you can present it as a proposal for seeking funds or investors.

2.   **Get a Nice Location:** Location of your shop, store or company is also very vital in starting up a cannabis business. Make sure the location and environment are weed friendly. Make sure you survey the area well enough before pitching your tent. A very unfriendly location is not good for business. It's either you meet with sabotage or very low patronage.

How friendly are the laws guiding cannabis involvement in that particular location? What are the costs you might incur in the location? These are some of the important questions you might want to ask yourself while searching for a nice place to start up your cannabis business.

3.   **Form a Business Structure: T**his is highly important before you are being granted a license. You would really need to show the cannabis world

that you didn't come here to play. Organize and draft the structure of your company. Point out the key areas and positions.

It's one of the necessary requirements a company must have in this line of business before you would be granted a license to start trading. You might want to get in touch with a lawyer if your company is going to be a huge one. But if you plan on starting out small, then there's really nothing much to do.

4. **Get the License:** Now, this is the centerpiece of the whole business. This is the legal key that gives you the right to open up a cannabis business. Remember, getting a license is not really as easy as walking in the park. It might take quite a long period of time. Thus, you would need to be very patient.

Additionally, it is advisable to go over your business plans and structures repeatedly, so as not to make a mistake that may end up costing you a license or even delay the process for a bit longer. The marijuana industry takes everything seriously, no matter how little or even irrelevant. So, endeavor not to make any mistakes.

5. **Start Getting Operational:** What are you waiting for? You have your license already; the next thing to do is start being productive. Start putting your plans into action and start making money for yourself. You would also need to follow every part of the regulatory measures as listed in your license in the cause of being operational.

Thus, everything you do should always be in accordance with the industry. Every decision you take should be inside the jurisdiction of the industry. That way, you would be able to enjoy a peaceful and productive reign in your business as you start making money out of it.

6. **Contracts and Partners:** As you grow in this line of business, you may want to expand and enlarge your outlets and reach. This is the greed that comes with every business. Well, as you plan on doing that, make sure you get the right people to partner with. It is important to know that the kind of business you intend to start highly determines the kind of partners you need.

As a skin care company whose major ingredient is the marijuana plant, you would want to partner with an advertising or marketing giant to boost your sales. A big shot cannabis farmer or grower might need to enter a contract with retailers and distributors.

7. **Promotions:** There is no business that can survive without promotion. And even if it did struggle to survive, it won't be for long. Promotion is the life of every business. How well you can promote your business determines the growth of your business. In case you are starting as a small-time businessman, you can utilize social media to your gain.

Advertise your products there. Place banners and posters around the city. You can also make use of trade journals as a cannabis cultivator. They are very useful and come in handy most of the time. There are lots of advertising forums where you can showcase your business. Make good use of them.

Becoming a successful businessman or woman comes with certain steps and decisions you must take. Sometimes, you need to be cutthroat. Other times, you need to let loose and go with the flow. This same step applies to the cannabis line of business. Now that you've known the compulsory steps to take, how about knowing about how to invest in the industry?

Chapter Ten

Investing in a Marijuana Growing Venture

Now that we have fully explained and examined how and why you should consider going into the canna-business, we are sure you would agree that this line of business is not totally bad as many had portrayed it to be. As a matter of fact, lots of self-righteous and upright men and women have joined this money-making industry.

Don't be too shocked when you find a top administrator, celebrities, and so much more becoming owners of a giant cannabis plantation. There is nothing to hold against them if you are still in the dark days of stereotypes. Instead, think of it as a business activity. They saw a money-making venture, and they jumped at it. It's that simple.

Be that as it may, if you still don't feel alright with the idea of owning or starting up a cannabis business of your own, then this chapter is basically for you. If you want to stay as a ghost or watching the whole thing play out from the back, then I'll advise you to read this attentively.

How about being an investor to a Canna-business? Sounds great, yeah? This is another way for you to make money and still not be involved in the play as you have always wanted. It's a perfect way to stay in the dark. All you need to do is just to throw some money at the business and watch how it's being run by another.

Just like buying shares, stocks, and bonds, this is also not quite different. You will be shocked at how much return this line of business would bring back within the shortest time. I'm sure I don't need to tell you about how the cannabis plant is winning the war against it in recent times.

With more than 25 states in the United States of America legalizing and removing the ban placed on this psychoactive herb, business has been booming ever since. All thanks to science, the plant is now being seen in a new light. Instead of the harmful substance it was earlier known to be, it has now become the savior of mankind. This is all thanks to medical marijuana.

With this new development, there has been an all-time increase in the number of investors in this newbie of an industry. It's now become clearer that the risks in this business are diminishing gradually. Thus, stamping their foot in this industry in the early stages would be a welcome idea. In the future, when the industry has reached its peak, this set of investors that took a leap of faith with the emerging industry would now be the top notch. This is a pure strategy.

For example, when the introduction of crude oil just broke into the American markets, Rockefeller, who happened to be one of the few people that took a leap of faith in this line of business during its early stages, made his mark afterward. He became successful and even topped the list of the richest men in America.

The same thing is applicable to the cannabis industry. As a newbie in the American economy, now is the best time for you to make your move. Now is the time for you to make sure you write your name in paint with this sector. Nevertheless, the success won't come easy, though. There would definitely be setbacks. Investing in this industry is a 50:50 chance. You have to take a chance on it.

Like we all know, all business and investment come with a few glitches and risks. The ones that don't have the risks are either well-grounded or protected. Now, the question you should ask yourself is, are you willing to take a chance on this new industry or a new business with no record of success and profit? If you are, then it shows you didn't come here to play. But if you aren't, then I guess you still need to do your homework well on the industry.

With that being said, this new industry has continued to progress and make new developments even with the series of jurisdiction, restriction, principles, and laws guiding the industry. In the United States, there are lots of laws which are binding and serve as checks and balances for this line of industry. The sale, distribution, and use of the cannabis plant have been quite strict with laws guiding it.

The introduction of medical marijuana has opened up a new path for lots of companies and investors that are interested in the line of this business. Many of them see this industry as the next big thing in the country. They believe it's only a matter of time before the industry will become one of the most profitable in American society.

Medical companies now see marijuana in a different light. These companies have made millions of dollars from the recent advancement and breakthrough of the plant in curing and becoming an alleviator to lots of deadly diseases out there. Its multipurpose medicinal properties can be said to be a blessing to mankind.

In Canada, there is an immense success rate of medical marijuana and this has further aggravated the advocates of the cannabis plant to start pushing for the legalization of the recreational use of the plant. The current situation of things looks like the government might give in to it sooner than we had anticipated.

This trend of events has pushed lots of people into investing in the industry. They believe Canna-business is the future. Its steady growth in Canada has opened up a pathway for these companies and investors to expand their reach into the United States.

Be that as it may, it is important to know that cannabis plant investment comes with a lot of risks which would serve as a direct negative impact on your investments. Experts have outlined this as one of the most downgrading blemishes in this sector. As much as the sector is quite amazing and filled with a high rate of profits, there is still a chuck load of uncertainties surrounding it.

The laws and principles guiding the industry today might change to something different tomorrow. The laws that favor your investments today might become detrimental to your investments tomorrow. That is how this line of business works. Now, the question is, are you ready for this? If you are, then check out our list of businesses you can easily invest in.

Canna-Businesses to Invest In

Over the years, lots of companies and investors have delved into making the best use out of the marijuana plant. With the introduction of the cannabis industry, there have been lots of opportunities which a series of wise companies and investors have utilized to produce different products with marijuana as a key ingredient.

From the local one-man store, down to a partnership company, they all involve themselves with cannabis either directly or indirectly. This industry has provided a tent for these various establishments to accommodate themselves and thrive together under one banner: marijuana. That way, investors can find it easy to invest their money in any marijuana-involved establishment of their choice.

Planning on investing your money in a cannabis establishment? Then here is the list of Cannabusiness that may interest you.

1. **Agricultural Technology:** You can invest your money in the innovation and invention of agricultural equipment and appliances that would aid the cultivation and harvest of the cannabis plant. For example, you can invest in establishments that produce sophisticated lighting, automated fertilizers, harvesting equipment, and so much more.

With the increasing grounds cannabis production is gaining each day, I can boldly say that this line of cannabis business would be much more profitable in the future than it is now. The more the plant gets legalized in a particular country, the more growers will need this equipment and technology.

2. **Supplementary Products and Services:** This is also a great business one can invest in. There are lots of companies out there that focus on providing the industry with supplementary products and services. These products and services can come as insurance packages from the insurance companies out there to the cannabis farmers or companies.

There are also lots of companies out there that offer packaging services or even promotion of the cannabis finished products. Investing in this type of business is not a bad idea. No need to get involved with the plant directly; just drop some money.

3. **Consulting Services:** This can also be known as the legal companies involved with licensing, paperwork, and zoning for clients. These clients can be a firm in the cannabis industry. If you are a lawyer, then investing in this line of business would be great. As a consulting service, you would be responsible for the maintenance of your client's operational processes.

4. **Consumption Devices:** This is within the arts and crafts division of the sector. Mostly, they are a one-man store business with little outreach, basically in a community. They produce consumption devices like pipes, hemp materials, cakes and pies laced with cannabis, and so much more. Invest your money in this business and be sure to expect great returns afterward.

5. **Retail Cultivators:** Sometimes, the growers of the cannabis plant end up doing business directly with the buyers instead of following a middle man - a retailer. This is mostly because of the maximization of profit or even breach in the agreement between both parties.

Nevertheless, imagine pouring your money at such cultivators. Without a middle man to leach them off their profits, the numbers in the returns would surely look great. With your investment, the growers might start expanding. More expansion can only mean one thing for you, more returns.

6. **Cannabis Products and Extracts:** A lot of companies out there focus on getting the best out of the marijuana plant via their sponsored research and experiments. They have been able to produce lots of finished products with the marijuana plant like drinks with marijuana content, products with cannabidiol, edibles, and so much more. Investing in this type of business is a win-win situation.

7. **Organic Farms:** Many people tend to prefer organic to chemical-based marijuana products. To this effect, lots of companies and individuals have taken it upon themselves to meet the demands of these people by creating and producing organic weed directly from their farms.

They sell these plants to companies or consumers who would want to make use of it either as they come raw or process them to any form they want. Invest in this type of business and your money is sure to come back in folds.

Risks That Comes With These Investments

Like we said above, risks are synonymous with any kind of business, especially the cannabis business. The same thing goes for investments. As a matter of fact, the higher the potential profit to be gotten from this cannabis business, the more the risk involved. Additionally, it is important to know the kind of risks and their gravity one might come up against in this line of business. That way, you would be able to prepare yourself when they come.

1. Success is not Guaranteed: With the cannabis industry, it's better for you not to count on it as the success is not entirely guaranteed. Though lots of people and investors have delved into this business as it promises great returns, still, it is still not entirely certain as regards the success rate in the present or in the future.

Little wonder why lots of businesses and investors are banking on the distribution and sale of the plant, instead of growing it in their own organic farms. They believe this is the new money-making venture. But the process is still under consideration. In Canada for example, many of these types of businesses have failed to see the light of the day even after establishment.

They ended up falling to the harsh and rigid buildup of the industry without even getting the privilege of teaming up with the right investors, some of which are not even considered legal as they end up not meeting the necessary criteria. To this effect, license and other paperwork were not being granted to them and as a result, they couldn't attract the right investors.

Like we pointed out above, the cannabis industry is a newbie in the economy of the United States or even Canada. No matter how well and guided their procedures and processes had promised to be for investors to come in, there would still be lots of risks lurking up in the shadows, waiting to attack the investors any time. Be that as it may, there is no safety net to put the investor's mind at ease about their investments. So, think twice before putting your money into the Canna-business.

2. **Government Regulations:** The government can be a bore at times with this line of business and investment. The already laid-down laws in the industry by the government might take another dimension. This can come as an idea to test new government policies and see how well they will work out.

For example, the government might come up with a whole new initiative as regards this industry. They may drop a law that affects branding or even the sale of raw cannabis. This would definitely affect the related stores and companies out there that deal with such. This would also affect the rate of returns and profit. And low returns automatically mean low value for your investments.

Additionally, new laws by the government may be targeted toward providing a great business environment for other people or companies that may have the urge to join in this multi-billion dollar industry. And do remember, an increased number of companies in a particular line of business is a competition. Prices will adjust and fall into shape, business models will certainly change, your customer base will definitely be resized, and the value for your investment will certainly be affected.

3. **Pricing and Taxation:** In every line of a legally-registered business, there is always an issue of tax and other forms of payment to pay. When this pricing and tax is becoming too much, they eat deep into your profits and make your efforts look weak. This is where the success of your company is not guaranteed.

Cannabis companies and farms that produce cannabis products (both finished and raw) end up placing a relatively affordable price on their products. This is to make their products quite affordable and relatively cheap to their customers out there. But when the government pricing and tax ends up becoming too high, the only option left is for the companies to increase their product value. That is the only way they can get good returns without reducing quality. This may affect the customer base, and in turn, would reduce the value of your investment.

4. **Expensive Share Prices:** Some owners of cannabis businesses tend to inflate their shares and stock prices in the bid to boost and grow their base. They tend to bleed the investors dry so as to get an edge and stamp their feet in the industry. Additionally, the prices of the cannabis company shares are largely based on the current state of their performance.

This could lead to a loss on the part of the investors. What if you end up losing money with the uncertainty surrounding the cannabis industry? What if the overpriced shares end up not making the much-expected returns you had earlier anticipated? That would be a loss on your part as an investor.

Be that as it may, some cannabis business tends to come out clean by announcing their intentions to increase the prices of their shares as a result of growing and expanding their reach. Thus, it is your choice to pour and throw your money at such kind of shares. It is your choice to take such risks as there is a possibility that your investment might not even increase in value.

5. **Operating Costs Can be Quite Expensive:** Running a company can be quite expensive, especially a cannabis business. If you intend to open up a very large and productive marijuana dispensary or farm, then you should be prepared for the large operating costs it comes with. The labor and capital it comes with are also substantial.

Thus, instead of companies saving more with low labor and a minimum level of labor, they tend to dig into the profit in order to construct and create more products, so as to meet up with demands of the consumers. The profit that ought to have built up the value of your shares would be used to revive and rebuild more operating systems to help grow the business.

Thus, it is your choice to invest in this business after reviewing the business plan and checking out the strategies employed by the business to pull profits. You should also check the necessary risks involved and the cost of its running. If you are cool with it all, then you should go for it. Nevertheless, you should never forget that there is no guarantee in this line of business when it comes to the success rate.

The marijuana industry might be a money-making industry, but I never said it's as easy as taking a walk in the park. For you to come out on top, you need to be at the top of your game. You need to know when to swerve and when to take the hit. That is what makes one a good investor in this line of business.

Chapter Eleven

Medical Marijuana (Traditional)

From the era of the Pharaohs down to the Ottomans, from the old Chinese empire down to the ancient Roman empires, the medical use of the cannabis plant had been in existence during this period as indicated in the records and archives kept in today's libraries. Traditional medical marijuana can also be said to be the peak of medicine in the ancient days. What seems like a curse in the present society had been a symbol of good health, prosperity, and religious beliefs.

Traditional medical marijuana had been quite effective in ancient eras as it was predominantly used to cure lots of diseases and illness like indigestion, cancer, inflammation, and so much more. Just imagine the great medical content inside the cannabis plant, and if those medical contents are fully revealed and utilized. There would be no bigger blessing to mankind than this psychoactive herb.

Little by little, the plant found its way into many kingdoms, empires, and tribes with its effective and efficient properties. While some smoked every part of it, others prefer to grind and dilute it with water before gulping it down. No matter how you end up taking it, it performs the same function.

There are ancient quotes about tribes who would gather around a campfire just to inhale the plant. They cut the leaves and buds if the plants and threw them in the fire. It would form a thick smoke which intoxicated the moment they inhaled. This gave them a certain high which energized and gave them a clearer mind afterward.

Now, this chapter will touch the very existence of the cannabis plant as an ancient medicinal alternative and support. How had these empires of antiquity really utilized the medicinal properties of the cannabis plant? What type of diseases and infections were cured with the use of the cannabis plant? These and many more are what this chapter is going to delve into.

The Ancient Chinese Empire

One can boldly say that the first old empire to open its eyes to the possibilities of medical marijuana is the ancient Chinese empire. Their involvement with the cannabis plant had been traced down to over thousands of years back. Before the medicinal use of the cannabis plant even came into limelight in this old Chinese empire, other medicinal means were used to care for the body and bring sanity and fitness to the body.

Tai Chi and Acupuncture were the only available ways at that period. Afterward, some series of herbs came into being but they aren't as effective and efficient as the cannabis plant. As a matter of fact, they aren't as documented and having as much importance as the cannabis plant during this era. The cannabis plant was highly recommended by the physicians of that era. It even became a household name in the ancient Chinese empire.

Traditional Chinese medicine had been a systemized way of meeting people's needs with medicine. It became a system which encouraged lots of developments and improvements toward the utilization and curing of diseases and infections. From dietary down to therapy, it was and still is very effective in the modern world. It helps you focus on having a sound mind. It helps you in harmonizing the body.

That was what Traditional Chinese medicine was all about before the introduction of cannabis into the system. Ever since, cannabis became the leading herb used in curing and alleviating illness and diseases.

To this effect, the Chinese name for cannabis is Ma or even Dama. According to a renowned Chinese botanist, by the name of Hui-lin Li, he explained the introduction and uses of the cannabis plant in old China. In his words,

"The use of cannabis in medicine was probably a very early development. Since ancient humans used hemp seed as food, it was quite natural for them to also discover the medicinal properties of the plant."

Additionally, an old Chinese pharmacopeia which is dated to be over 2,000 years old (100 AD) had further shed more light on the relationship between the old Chinese empire and medical marijuana. It introduced cannabis as,

"The flowers when they burst (when the pollen is scattered) are called [mafen] or [mabo]. The best time for gathering is the seventh day of the seventh month. The seeds are gathered in the ninth month. The seeds which have entered the soil are injurious to man. It grows in [Taishan] (in [Shandong] ...). The flowers, the fruit (seed) and the leaves are official. The leaves and the fruit are said to be poisonous, but not the flowers and the kernels of the seeds."

You will be further shocked to see that the cannabis plant had been very useful to the old Chinese empire. No doubt, Hua Tuo (140-208) can be said to be the first person to ever hold a record of using the cannabis plant as something much more than just smoking. As a surgeon, he needed a quick form of anesthetic which would be used on his patients before performing any surgery on them.

Thus, he resorted to the cannabis plant which was gaining momentum at that period. And guess what? It worked just fine, making him the first person to ever use the cannabis plant as an anesthetic. So how did he do this? It's pretty simple. He dried the plants and ground them to powder. Afterward, he mixed the powder with wine before administering.

In the same vein, Elizabeth Wayland Barber who had been doing research in this line of study had further expanded on the "knock-out" properties of the cannabis plant. She agreed that the pieces of evidence found in the old Chinese books on the marijuana plant put a lid on the bottle of knowledge as regards the narcotic features of the plant.

Believe it or not, the world could have further advanced in the medicinal field than it is currently. With a series of setbacks and campaigns geared toward the criminalization of this high medicinal herb, it's only a pity for one to see the damage those campaigns have caused in today's world. Just like the old Chinese empire, we could have prevented and made a cure for lots of illness and diseases out there.

With that being said, Frank Dikötter, who is also a Dutch sinologist, had also given a concise and well-detailed explanation on the medicinal uses of the cannabis plant in the old Chinese empire. In his words,

"The medical uses were highlighted in a pharmacopoeia of the Tang, which prescribed the root of the plant to remove a blood clot, while the juice from the leaves could be ingested to combat tapeworm. The seeds of cannabis, reduced to powder and mixed with rice wine, were recommended in various other (materia medica) against several ailments, ranging from constipation to hair loss. The Ming dynasty Mingyi bielu provided detailed instructions about the harvesting of the heads of the cannabis sativa plant (mafen, mabo), while the few authors who acknowledged hemp in various pharmacopoeias seemed to agree that the resinous female flowering heads were the source of dreams and revelations. After copious consumption, according to the ancient Shennong bencao jing, one could see demons and walk like a madman, even becoming 'in touch with the spirits' over time. Other medical writers warned that ghosts could be seen after ingesting a potion based on raw seeds blended with calamus and podophyllum (gui jiu)."

The above quotation is the best explanation one can ever find as regards the old Chinese empire in relation to the cannabis plant. Be that as it may, while the modern world lays emphasis and focus on the top part of the cannabis plant, the old Chinese empire made sure they utilize every part of the plant in meeting their medical needs. From the buds down to the roots, every part was very useful to them. They boil the plants with water, smoke it directly, or grind it till it forms a powder or a paste.

The Old Egyptian Kingdom

The Old Egypt, like any other kingdoms and empires of that era also didn't take a seat at the back while watching their neighbors enjoy the benefits of the plants alone. As a matter of fact, they took the front runner as the cannabis plant became a part of their history, religion, and even culture.

According to history, the cannabis plant had found its way into the old Egypt as early as 2000 BC. Egypt as an old kingdom didn't care much about documentation of the cannabis plant, thus there is not much ancient writings or words about the early use of the cannabis plant for medicinal purposes.

But, a thousand years ago, the cannabis plant had been used immensely for a number of activities and uses aside medicine. It had been used for making ropes, clothes, oil, and so much more. Later on, it started its reign as the most effective and efficient drug in the old Egyptian kingdom.

Little wonder why Ebers Papyrus (1550 BC) mentioned cannabis as a psychoactive herb which had been used to treat and alleviate illness and diseases. It further described the cannabis plant as a medicinal substance which was quite useful in old Egypt. Aside from this, there are quite a number of old Egyptian papyri which talked about the medical benefits of the cannabis plant.

They are the Ramesseum III Papyrus (1700 BC), the Berlin Papyrus (1300 BC) and the Chester Beatty Medical Papyrus VI (1300 BC). Now, let's focus on the medical benefits of the cannabis plant on the old Egyptian kingdom, shall we? The cannabis plant was basically used in reducing the effects and pains of people suffering from hemorrhoids.

Additionally, it was further used in treating sore eyes in old Egypt. As of 2000 BC, the cannabis plant had been used to treat diseases and infections like glaucoma, cataracts, vaginal bleeding, and to some extent, cancer. Now, imagine if the cannabis plant medicinal properties are fully harnessed and utilized; there would definitely be no limit to the medical solutions the cannabis plant would offer.

The Old Indian Empire

The old Indian empire also had its own fair share of the cannabis plant. As a matter of fact, the cannabis plant went far beyond medicinal usage with this empire. It became a symbol of religion. According to the legend, Lord Shiva had eaten and referred to the plant as part of him. The plant refreshed his mind and gave him contentment. Thus, it became more than just an herb for the worshippers of Lord Shiva.

Additionally, there is a text from an old collection which had laid emphasis on the existence and use of the cannabis plant for medicinal and religious purposes in the old Indian empire. The old texts revealed the psychoactive nature of the plant and the physicians and doctors of that era made good use of it in curing illness and diseases.

Insomnia, headache, gastrointestinal disorder, severe pain, and many other deadly diseases were eradicated or simply alleviated by the cannabis plant. The name given to cannabis in the old Indian empire was Bhang. Ever since then, quite a lot of ancient physicians, botanists, and philosophers had laid emphasis on the multidimensional purposes of the plant.

While some envisaged that the cannabis plant was the combination of both religious conscience and ailment control, others argued that the cannabis plant was about goodwill, happiness, and prosperity. They believed cannabis meant prosperity and good living. If Lord Shiva himself could use the plant, then the plant is surely a heavenly plant.

Other ailments the plant cured were dysentery, sunstroke, clear phlegm, quickens digestion, and so much more. Even the Old Indian empire agrees with the fact that the cannabis plant gives a clear mind, boost alertness, sharpens the thinking faculty, reduces pains in the general body, keeps the body fit, and connects the body and mind.

That is how much importance the plant was to the people of that era. Both Sushruta Samhita (600 BCE) and Vangasena (the writer of the old Indian text, Gerrit Jan Meulenbeld, Dominik Wujastyk, and Chika-Sara-sangraha) agree that the marijuana plant is not just a plant; it is the savior of mankind. They both believe the cannabis plant treats catarrh, phlegm, and diarrhea. It is also an appetizer which can be taken before eating.

Many old Indian spiritual leaders had also given their opinion about the uses of the cannabis plant. They believed the cannabis plant usage goes beyond its medicinal properties. It can also be used as an ingredient for religious activities. For example, the smoke that erupts out of the cannabis plant can be used to dispossess one's enemy and it performs its functions in the human body as fast as possible.

In some old scriptures like the Dhanvantari Nighantu, Sarngadhara Samhita, and Kayyadeva Nighantu, the cannabis plant was also painted to be blessed with high medicinal content. The scripture is believed that the plant contains traces of pain relief and an aphrodisiac. It also made mention of the addictive nature of the users. The consumption of the plant over a long period of time would definitely lead to addiction. Thus, it should be smoked carefully as it could end up damaging the lungs and liver.

Gradually, the cannabis plant found its way into old India from the ancient Chinese empire and started gaining momentum almost immediately. For thousands of years, after it came into India, it had established its base deep into society as one of the most beneficial plants with immense qualities. Further research had also shown that the medicinal use of marijuana in the ancient Indian empire dates back to the 9th and 10th century.

It was seen as the elixir of life. Cannabis became the solace many broken men and women could turn to. They now found a new friend in the plant. They found something that would make them feel light-headed, something that would make them feel amazingly great aside treating their diseased body. This was what the cannabis plant was to the people of this era.

The first ever clinical tests on the medicinal benefits of the cannabis plant were carried out by William O'Shaughnessy who was a British doctor during the colonial rule in India as of 1839. His findings became the very first research based on the introduction of the cannabis plant to the world of medicine. His effort and work didn't go unnoticed as he was blessed with a royal title for making an immense contribution to the development and improvement of natural knowledge.

Ever since William O'Shaughnessy had unlocked the immense benefits of the cannabis plant to mankind in relation to medicine, it had stayed legal with the young generation of medical students encouraged to build and direct their knowledge towards this line of medicine. Additionally, Ayurvedic medicine had become integral in the Indian medical society of today.

The Old Greek-Roman Empire

The old Greek-Roman era was filled with lots of exhilarating events, activities, and inventions that had made the people of this era stand out throughout history. To say the Greek-Romans had not come in contact with the cannabis plant would be an understatement. As a matter of fact, they had their own share of the plant. The ancient Greeks used the plant to treat and heal infected wounds and the seeds to kill tapeworms.

It was also used to treat inflammation and pain during the era. When the Greek soldiers left for war, they made sure they went with the cannabis plant as it served as pain relievers and alleviators. Herodotus who is a 5th-century Greek historian wrote some passages on the uses of the cannabis plant. The cannabis plant according to him may attack illness and disease, but how they are applied can be very different.

While some people grind, pound, blend, or even soak the plant, others just dry them and smoke every THC out of the plant. For example, the Scythians included the cannabis plant in their steam bath. In the course of history, Discordes, who was a physician in the 70 AD in Greece, was appointed by the Romans to research on a selection of herbs. To this effect, a book was created out of this research titled "Meteria Medica".

The book contains lots of knowledge about diverse herbs with descriptions and details. To sum it up, Discordes made sure he researched over 600 plants and these included the cannabis plant. During this period, the plant was used for purposes other than the medical benefits. It was used for ropes, clothes, oil, and so much more. The research of Discordes was highly successful and recommended by the nobles of Greece and Rome.

It was translated into so many languages as it was one of the few evidence of the medical properties and uses of the plants. Since, it was used as a reference for today's medical marijuana. A lot of people didn't use this phrase, "all roads lead to Rome," for nothing. So, could it have been possible for Rome to be the first empire to discover this psychoactive herb?

Galen, who was also the most experienced and popular physician and surgeon in both Rome and Greece, also gave his own opinion on the marijuana plant. He had emphasized on the flowers of the cannabis plant. To his understanding, the cannabis flower could be very effective as it would induce good feeling and extremely happy moments.

Additionally, Ovid, who was also a well-renowned poet during this era, had written lines about the cannabis plant. In one of his poems, he had told a tale of Glaucus. He had described Glaucus as someone who found himself eating a particular plant with a palm shape. He went further to explain how this plant made Glaucus feel happy and ecstatic. Even a dumb person would know that Ovid was referring to the cannabis plant.

Would it have been correct if I was to say that the ancient people knew and utilized the medical benefits of the cannabis plant more than us? Inasmuch as that may sound like a slap in our faces, especially with the tons of sophisticated machines and equipment, some researchers believe it's true. Now, how advanced is the modern day medical marijuana? Our next chapter will fill you in with all the necessary details.

Chapter Twelve

Medical Marijuana (Modern)

The previous chapter emphasized traditional medical marijuana and how effective it was for these empires and kingdoms. Right from the inception, this had been the primary use of the cannabis plant. It is widely believed that the old kingdoms and empires still possess the total utilization of the medical benefits of the plants, unlike the recent medical marijuana episode that is still young and just having its fresh series of breakthroughs.

Medical marijuana started as a result of the need for something much more efficient and effective, something that would combat and fight the hundreds of deadly diseases out there and keep mankind even healthier. Ripple by ripple, medical marijuana is opening the eyes of modern society to its outstanding benefits and qualities.

It has moved to pass the recreational phase and even developed into something much more. It had become a powerful source of medicine in the field of medicine and pharmaceuticals. However, this had seemed not to be enough. Some people still exist with the mindset of stereotyping and relating the plant as the devil's plant. Irrespective of the immense benefits and advantages this plant has showcased over the years, they seem not to be moved or swayed.

This is a result of the awful picture the government and other capitalist giants have painted of the cannabis plant. Thus, no matter how hard the advocates of this plant manage to scrub the stains off the image of the plant, there will still be traces left. In other words, there will still be people (traditional) who will still stay glued to their beliefs of the cannabis plant, irrespective of how many lives the cannabis plant tend to save.

Be that as it may, modern day medical marijuana focuses on managing the psychological and physiological parts of the body. It pushes away various antibodies that tend to live off the body, thereby, hurting the body in the process. There is no better way to fight off deadly diseases in this modern world than the application of modern-day medical marijuana pills and medicine. Over and over again, it has proven to be a lifesaver.

Little wonder why there have been quite a number of private or public funded researches and studies on the subject matter of marijuana. Scientists and researchers have tested and made sure the plant undergoes various experiments as regards the creation of alternatives and panaceas to the majority of the deadly diseases out there.

This is why I will urge you to drop the stereotypes and old merciless perception of the plant. Stop seeing the plant as a harmful substance or a deviant, hyperactive drug. Instead, start seeing the drug for the good work it has been used for thus far. Start seeing the cannabis plant in a whole new light where it has benefited mankind and helped provided a solution to the ailments out there.

The misunderstanding and misconception of the plant can be directed toward baseless assumptions and unrelated facts which had been propagated via the media houses. These cynical forces had their aims and objectives as they continued pushing towards the criminalization of the plant. As a matter of fact, they succeeded in brainwashing and clouding the judgments of the general public to a very large extent.

This has been seriously curbed to the minimum with the recent improvement and development of modern-day medical marijuana. We can actually say that the 19th century in the history of the United States witnessed a serious use of the cannabis plant in treating diseased patients. Physicians of that era gladly prescribe the cannabis plant as a drug in combating various ailments.

Diseases like nervous disorders, labor pains, nausea, asthma, sexual dysfunction, and headache were mostly discovered to be remedied by the marijuana intake. Thus, the prescription of the drug became quite popular in this century. This led to the surge in the research of the cannabis plant, its outstanding properties, and its usage.

Over a hundred studies were undergone and generally published to the community as regards the effectiveness, efficiency, and validity. It was the savior of mankind during this era as a lot of people believed in the efficacy of the plant. It was generally and publicly available and cultivated in the 19th century. In fact, historians had established the fact that in the history of the cannabis plant, the 19th century marked the most generally acceptable era of the cannabis plant. Some even went ahead to say that the cannabis plant became the smile on the faces of the people.

However, this smile wasn't meant to last forever. In the year 1937, everything changed about the plant. The churns of propaganda, stereotypes, and ridiculous campaigns were put into place. The plan was to defame and demoralize the cannabis plant - and guess what? The plan worked perfectly well. The once-celebrated plant became a poison that must be flushed out of the system and society at large.

This gave birth to the Marijuana Tax Act of 1937. This law never recognized medical marijuana to an extent and the rest of its uses are seriously prohibited and seen as a crime against the government. The law believed that the medical properties of the plant were a myth that could not be proven, thus, subjecting the plant to a mere deviant and menace in the society. It sees the plant as a harmful substance that would not only harm the old but the young and future generations, also.

They started formulating a series of rumors and unfounded stories about people going crazy as a result of taking the cannabis plant. Within the shortest time, it got the support of the people as everyone started seeing cannabis users as people prone to wild behaviors and psychological disorders. This is not only wrong, but also immature. These are absurd stories with no fact, but to a very large extent, it managed to win the public favor.

As Preston Peet puts it,

"The underlying purpose of criminalizing and vilifying cannabis was to kill the hemp industry and make the big business interests who ran the synthetic fiber industry even more filthy rich. It seems possible that the alcohol industry, which had just exploded after the end of prohibition in 1933, also stood to benefit from a total ban on marijuana."

Ever since, the cannabis plant stayed in stealth and very clear from the outside world. Only those who have the passion and courage for it had been able to grow the plant and keep the plant from being extinct. The medical world stayed completely clean from the cannabis plant. It no longer serves as a drug which can be prescribed to patients.

Nevertheless, as the saying goes, what goes up must surely come down. Same goes for the negative stereotypes and propaganda web spun around the cannabis plant. The new world started seeing the plant in a whole new light and the stiff mood revolving around the plant started getting milder in the early 1990s.

First among its peers, California broke the jinx by passing Proposition 215 into law. This would enable the state to operate freely on medical marijuana. The people could now get access to marijuana-motivated drugs to treat their ailments. This single act had opened up the eyes of many other states in the United States. Gradually, they had also started opening up to the reality of things.

As of 2017, over 25 states of the country have legalized medical marijuana, and ever since, an immense improvement and development has been recorded in the field of medicine. It has proven to be one of the wisest decisions ever made by these states since the beginning of a new millennium.

Medical marijuana is now a very important part of modern day medicine with doctors actively and free to prescribe the cannabis-innovated drugs to patients if the need arises. So far, the cannabis plant has been able to cure or alleviate various ailments like vomiting, migraines, multiple sclerosis, glaucoma, nausea, sickle-cell disease, Parkinson's disease, sleep apnea, spinal cord injuries, arthritis, Alzheimer's disease, muscle spasms, and asthma. It has also been able to cure and control various psychological conditions like anxiety disorder, insomnia, bipolar disorder, depression, and PTSD.

Additionally, there has been a lot of research and studies carried out to further push and invent more remedies and panaceas to the deadly diseases out there. This has so far been very consistent and fruitful with lots of improvement and development being added to this line of medicine. For example, the new research carried out and tested on laboratory animals has proven that the lists of Cannabinoids stored in the cannabis plant can be used to stop or even kill off a tumor. The tumor needs the blood vessels to grow and this is what the cannabis plant is going to cut off.

This might be the permanent solution to stopping cancer and saving thousands of lives. This would make the best cancer treatment at the moment (Chemotherapy) obsolete. Instead of killing off both cancerous cells and the healthy ones like Chemotherapy, the cannabis plant would only focus on the cancerous cells. This is an advantage over Chemotherapy.

According to the University of Nottingham, the cannabis plant is also showing its strength as regards neurological functions. What does this mean? It simply means that the cannabis plant has the tendency to move a part of the body that had been paralyzed after a stroke. In another study, the cannabis plant made a significant improvement toward the protection of the body from getting attacked by further cancerous cells. In other words, it stopped the development of cancer in the body and stops the cancerous cells from developing more tumors.

Well, why not let's make fun out of the marijuana story? In a world filled with love and affection, a strange but amazing fellow walked into a beautiful place filled with lots of kind-hearted people. After a while of establishing himself as a trustworthy and wonderful fellow, the people trusted him and started relating with him. It got to a point that they now placed their trust in him.

With his multi-purpose and multidimensional functions, he became the physician that soothes the people's pain away. He would cure them of their illness. He also became their priest,;he would listen to their confession and relax their minds by giving them a clear head. He became their motivation. He will give them inspiration. Tell them stories of how great men were made and how some fell. It was their choice to choose which they would want to become.

Then, the elders of the land got jealous. They used to be the pillars of the land, the foundations, and people loved them. But now they had lost the love and adoration to a novice who just entered the community. Something had to be done! This is where the lies and betrayal came into play. The elders began to churn up stories about the young fellow who was only there to help.

They brainwashed the people until they chased and crucified him with their own hands. They had forgotten the help he had rendered. They had forgotten how he kept their secrets and showed them a new path. They had forgotten how he had saved them more than a thousand times without asking for a dime in return. Just as he came into the community, he also left. But not without leaving sweet memories in the minds of a few others.

These few others started fighting the injustice on behalf of the fellow. At first, it wasn't easy, but the good memories they had kept them going. After what seemed like a long fight, they won the fight. They cleared his name. It was now time for the fellow to come back and reclaim his lost reputation. He knew it wasn't going to be easy, but he was willing to try.

With welcome hands and sorry faces, the people of the community were happy to be reunited with their savior. They knew they had wronged him by not believing his story. However, they knew everything happens for a reason. Same reason why the new relationship they were about to build would stand the test of time. Amazing story, yeah?

Presently, the fame of medical marijuana has gone far and wide with lots of government and nongovernmental health associations joining the train. We can proudly say that over 50 United States and other international health organizations have endorsed the use if the cannabis plant on patients that need it. However, it must be strictly under the supervision of a physician experienced in the field of medical marijuana.

Some organizations have even gone ahead to endorse the implementation and creation of medical marijuana clinics on a wide scale. That way, both medical practitioners and patients can really have access to unlimited medical facilities and services.

Modern Day Breakthroughs

1. **Childhood Epilepsy now has a cure:** This was one of the best news in the year 2018 with the newly established breakthrough of the cannabis plant in the world of medicine. With this, I'm sure those with negative perceptions about the plant will finally begin to see reason on why the plant should be celebrated as the savior of mankind and not poison.

To this effect, Epidiolex was created out of the cannabis plant to fight epilepsy in children. This drug would combat and fight seizures in children which are caused by Lennox-Gastaut syndrome or Dravet Syndrome. These are two disorders that can be inherited and passed down from generations to generations.

What are the contents of this drug? It is a pure and undiluted drug which was created from cannabidiol. The drug was developed by GW Pharmaceuticals which was as a result of the touching story of Charlotte Figi, a little girl who suffered from the hands of Dravet syndrome.

Her story made the news as of 2013 when her parents told the world about how they had tried every possible solution from drugs, down to dietary and surgery. All these were directed toward controlling the regular seizures that had been disturbing the young girl since she turned five months old.

However, when other drugs and remedies seemed not to be fruitful, medical marijuana was. Her parents shifted their focus from normal drugs and focused their attention on medical marijuana. There was an immense improvement in the health of Charlotte afterward.

Her seizures stopped entirely with the help of medical marijuana. This sparked the recent breakthrough on the aspect of childhood epilepsy. A lot of studies and research was carried out and the result has been quite encouraging. A study published in the New England Journal of Medicine also gained momentum in 2017. The research further proves how cannabidiol can be used to curb this ailment in children.

Since the introduction of Epidiolex, there have been quite a number of improvements in the management in the area of childhood epilepsy. The seizure is stopped considerably, giving the lives of these children a whole new meaning. The drug has further made child epileptic seizures become very ordinary, like catching cold or everyday headaches.

2. **Migraines can now be prevented:** I have seen lots of people suffer from this painful and head-pounding headache and trust me, it wasn't a beautiful sight to see. Many of the treatments out there are just not working at all or are being minimal in their effectiveness. How about you try a new way? How about you try an effective drug this time around?

Aimovig is a drug created from the cannabis plant. How does it work? It simply stops the functioning of the molecule which causes this throbbing headache. That way, the pains and effect of migraine would not be felt and in no time, you will feel better, like you never had any migraines to begin with.

According to statistics, over 10% of the world's population suffers from this annoying ailment, and more than half of the population are women. Thus, Amgen Inc saw the need to enter into the market by creating a drug that would not only minimize the pain but also eradicate the ailment once and for all.

In the cause of creating and perfecting the Aimovig, there had been lots of trials and tests along the way. It was discovered that this drug has the tendency to reduce migraine a whole faster than any other drugs out there. Additionally, this comes with very little or no side effects. That is good news on the part of the users.

Thus, Migraines can now be prevented and even eradicated faster than ever before. And believe me when I say there is going to be an improvement on this drug. It's only a matter of time before it happens.

3. **Non-Opioid Withdrawal Drug:** Opioid eradication had been one of the leading campaigns the United States government has championed over the last few years to date. As a matter of fact, it has become really popular with the number of deaths from opioid going on the rise every year. In 2016 alone, the statistics show that over 100 people died daily from an overdose of the drug. That was really sad.

I'm very sure a lot of people into this harmful substance would want to desist from harming themselves but are very scared of the withdrawal symptoms which would engulf them almost immediately. There is no benefit oxycodone, fentanyl, codeine, and heroin give to the body. As a matter of fact, it keeps taking from our body until it will take no more. It will take our health, our blood, our mind, our soul, and our sanity.

An attempt to stop and reorganize our lives would bring us nothing but more pain as our body starts getting symptoms of discomfort like unnecessary fear, anxiety, nausea, and sleeping problems that would make us feel really uncomfortable until we get a taste of the opioid we had promised to leave behind. It's that crazy.

Before the development of a real solution, there had been a procedure which was put in place for the gradual reduction of opioid addiction. Here, the doses of the opioid will be reduced gradually or even swapped with less harmful and strong drugs. Though it had worked, there had been an argument against this procedure.

Today, Lucemyra which is marketed and distributed by the United States WorldMeds, has proven the traditional procedure as obsolete. It became the first ever non-opioid withdrawal drug that has proven to be very effective. The drug works in such a way that it helps reduce the neurochemical in the body. This chemical is what makes it difficult for the addict to withdraw permanently from the use of an opioid. It has proven to be a help, instead of an alternative to opioid-like the traditional procedure.

4. Most Potent HIV Medication to Date: The Biktarvy, which was created by Gilead Sciences, has proven to be the most effective drug derived from the cannabis plant to combat HIV. The virus is no longer seen as a no cure ailment as it was made to be in its early years of inception.

Though there is no direct and permanent cure to the virus, there is still a whole load of antivirals which would help keep the virus in a safe stage. Now, how does Biktarvy come in? As a 3 in 1 pill, it should be consumed or swallowed by the user, so as to serve as a mechanism that prevents the virus from going overboard.

The Biktarvy is made up of bictegravir, whose only function is to fight and barricade the HIV from attaching itself with the host DNA. That way, the person can stay free from the virus advancing into its AIDS status. A lot of trials and tests had been carried out with the drug before administering it on humans and it has proven to be effective and efficient ever since.

These and many more have been the regular breakthroughs the cannabis plant is being known for. Year in year out, there is always a great discovery in the line of medical marijuana. The developed and improved medical system had made it possible for new research and studies to be carried out. Thus, making medical marijuana a blessing and not a curse. This brings us the beginning of a new chapter; you won't want to miss it.

Chapter Thirteen

Marijuana (As a Pain Reliever)

In the last chapter, we discussed mainly the face of medical marijuana in today's world. We also shed light on some of the massive breakthroughs that had been made as regards this field of medicine. These feats are no small feats as they have further made the world become a much better place.

For example, in the 90s, HIV & AIDS could be agreed to be an ultimate death sentence with the virus spreading across the globe faster than ever. Within the shortest time, it had crossed into all the continents of the world. This gave birth to the tons of research and studies geared towards the reduction of the effectiveness of the virus, at least if not the complete eradication.

Alas, the long-awaited Messiah the people had been waiting for as regards curbing and eradicating deadly diseases like HIV and cancer had been right there all along. We just didn't look hard enough - or maybe we just weren't ready. In early 2000, medical marijuana entered the line of medicine and since then became the leading phenomenon. It has provided lots of solutions and panaceas for lots of deadly diseases out there.

Aside from these diseases, marijuana has also become an alleviator for chronic pain. Instead of just applying the traditional procedure in staying off pain like medications and opioids, why not just make use of the marijuana plant? In the old and ancient kingdoms, particularly the Chinese empire, it has been recorded that physicians of that era had not only made use of the plant for recreational and medical purposes but to also alleviate and relieve their patients from pain.

Marijuana makes sure it takes the pain away swiftly and easily without leaving any side effects. The types of chronic pain the marijuana plant alleviates include the nerve damage pains and pains from inflammation.

Additionally, chronic pain extends to even people suffering from more deadly diseases like cancer and diabetes.

The results of a long term chronic pain without having a solution to either cure it or suppress it can be very dangerous and detrimental. It often leads to disability. As a matter of fact, statistics have shown that the number one causal agent of disabilities in the United States is the chronic pain people experience.

It is important to know that the United States government has still not come to terms with the legalization of the cannabis plant. Though more than half of states in the country have agreed to lay down their guard and further embrace medical marijuana, this action still didn't sway the federal government into changing its decisions.

This is why the Food and Drug Administration (FDA) of the United States still doesn't recognize the use of pure marijuana in combating chronic pain. According to them, it is too risky and cannot be trusted. They believe there has not been enough substantial evidence and fact to ensure the effectiveness, efficiency, and safety of the use of marijuana as a panacea to chronic pain.

Nevertheless, some other bodies had argued otherwise. As a direct opposite of what the FDA believed, they have agreed that the marijuana plant used as a remedy to solve the chronic pain ailment is very effective and efficient, as well as safe. According to them, anecdotal evidence which showcases how the cannabis plant compounds would be of help in reducing and alleviating chronic pain should be enough to endorse the marijuana plant.

Chronic Pain

With that being said, it is important to know that the type of chronic pain you are trying to alleviate signals the kind of cannabis strain one should use. Pain is the first thing we feel when our body system is being attacked by outer or foreign bodies. In other words, it is an alarm which gives a signal to other parts of the body when all is not well with us.

If you have someone pinch you hard on your skin, you will definitely make a loud noise as he or she pinches. The more you feel the pinch, the louder your voice would be. This is how pain works. The more pain we feel, the louder our screams would be. This would now tell those around us that we need urgent medical attention.

Sometimes, our screams from pain might be a forced alarm. It might not be a serious problem or ailment at the end of the day. In other words, pain has more than one cause. And before we can alleviate a certain pain we are feeling, we really need to ask ourselves these three important questions first:

1. What is the pain?
2. Where are we feeling the pain?
3. And why are we feeling the pain?

The answers to these three questions would aid us in knowing exactly how to reach our pain and eradicate it. Some pains come from emotional distress or even psychological situations. For example, there is no drug or medicine for heartbreaks. Instead, you can find a therapist to walk you through the pain. But that doesn't reduce the pain you will feel or even divide the pain between yourself and your therapist.

Thus, there is a need to develop and come up with a whole new drug that would encompass all forms of pain. A new drug that will ensure that what you are feeling doesn't get blown out of proportion. To this effect, there was a need to go back to the early panacea that had managed chronic pain well but was dumped into the trashcan of history.

When people started going back and finding real solace in the arms of this drug, it has been further tested and particularly used in eradicating pain incurred situations and conditions like headaches, inflammation, migraine, pains felt after or before childbirth, and so much more. A lot of public symposiums and speeches were held with advocates and people with testimonies dishing out the wonders of this plant as regards chronic pain.

The marijuana plant has since been held in high esteem amongst people suffering from this ailment. It has become the first solution they would refer victims of this ailment to whenever they have the chance. This marked the beginning of a new era for the marijuana plant aside from the renowned feat it had made as regards medical marijuana.

How Do We Feel Pain?

Now, if pain is the alarm of our inconvenience and uncomfortable moments, how then do we feel it? Where does it emanate from? We should know that our brain anchors the pain that we feel throughout the whole body. For example, if we are stung by a bee on our cheeks, the nerve sends in an immediate signal to our cells which in turn sends a direct signal to the brain. It is now the brain that will send us images and flashes of where we got stung. Quite amazing, yeah?

Pain reaches the brain in three different forms. This form depends on the heaviness and amount of pain. They may be acute, short and intense, or chronic. The acute pains are pains that occur when we just go through surgery of any kind. The kind of pain we feel during this moment is as a result of our body recouping to the way it was before the surgery. Medical professionals do recommend opiate or narcotic substances to help ease the pain.

Short or intense kind of pain is mostly short-lived. They sometimes end the very moment they begin. Sometimes, they don't take much time before the person starts healing up. Normal and regular soothing words are more than enough to cure this kind of pain. There is really no need to going narcotics. In fact, that would be totally out of line.

Lastly, chronic pain is more deadly. They may end up leaving the person unfulfilled, unhappy, and even consider ending it all. Opiate or narcotic use during this phase of pain would be highly inadvisable, because instead of helping out, they would most likely leave you in a much more ugly state. The side effects they come with it are very bad and hardly revocable. Thus, instead of harming yourself with solutions that wouldn't be of any help, why not think outside the box?

Opiate and narcotic substances might give you a temporary solution for a few months, but what if those months' elapsed and the side effects begin to set in. Would you be up for it? Would you be able to contain it? Do you think you will be able to manage it? This is where marijuana comes in. The marijuana plant would give you a very good solution to this pain without negative side effects.

Marijuana & Pain

According to the ancient books, pieces of evidence have shown us that the medicinal properties of the cannabis plant were discovered and made use of as at 2700 BC. Ever since, there have been lots of improvement and development of the medicinal use of the plant, even with the crude and unlimited tools and skills of the ancient people.

Even at that, they still managed to pull something out of the whole mixed up properties of the plant. The first person to ever find and record the analgesic properties of the plant was the old Chinese emperor himself, Shen Nung. Shen Nung was known to be a little obsessed with old medicine. Even as a young boy, he would memorize and read texts of medicinal plants. This can be said to be what had pushed him into delving into medicine even as an emperor.

To a very large extent, he was considered to be the father, founder, and originator of Chinese medicine. He even took it upon himself that his people didn't fall short of medical attention. That way, he would diagnose and prescribe herbs for patients himself, even as the emperor of the empire.

His Chinese medicine in today's world can still be said to be very effective and efficient. Though it might not be popular as it used to be, it is still being applied in its original version in some cases. While it has undergone lots of reformation and improvement, Chinese medicine has been very vital in the breakthroughs of medical marijuana today.

Right from the inception, the list of Cannabinoids present in the marijuana plant has shown a significant relationship with pain. According to the series of experiments carried out, the marijuana plant has shown great promise towards the eradicating and curbing of pain in humans.

Remember the nerves we mentioned above as a pain detector and transmitter in the body? They showed immense relation to the Cannabinoids present to the cannabis plant. What the Cannabinoids do is to ensure that the signal these series of nerves send to the brain is being cut off. That way, there would be no pain felt so long the brain didn't pick up any signal from the nerves.

Additionally, the recent study in opiates and the cannabis plant has further shown that both substances follow different channels in performing their functions as regards pain. In other words, they both make use of a different mechanism in exerting their influence in the bid to suppress pain. Thus, there is a strong chance for both substances to work perfectly together for even better and faster results. The side effects would surely be minimal as both substances would be tweaked to complement each other.

Be that as it may, it is important to know that these pain suppressing experiments using the cannabis plant have not really been officially carried out, especially with the rising problems emanating from both technical and ethical difficulties. The FDA, on one hand, was becoming a bone in the neck, while other private organizations are also there churning up obstacles on the road of this significant breakthrough.

As a matter of fact, the first and only study carried out in this line was in 1981. Ever since, every further attempt to start the chain of experiments on this line of the study has proved abortive. The nearest thing we could call an experiment was when some people personally tried alleviating their pain from cancer or surgeries.

These side experiments can be said to be of no use with just very few of them making use of an appropriate scientific method. Nevertheless, the advocates of this field of study have refused to be swayed and still maintained their stand on the marijuana plant being an active alleviator of pain. Their arguments relied heavily on the existing literature on the connection between the plant and pain.

Ever since there have been studies as regards using the cannabis plant as a pain reliever. The studies also prove that the drug possesses a directly opposite effect which many do not know. These effects are severe shocks, heat, and increased sensitivity to actual pain. One would see that the same cannabis plant that had been praised for being the suppressor of all pains now causes side effects that are even more severe.

In another study, it was proven that the Cannabinoids had failed to suppress the pain felt by electric seizures. But critics had criticized the study from two different perspectives. Firstly, the focus was given more to the responses to the pain instead of the sensations of the pain. Secondly, while the participants were subjected to just seizures of pain, they were now asked to take note of the pain they felt and its intensity. This is absolutely unacceptable.

Best Marijuana Strains for Chronic Pain

It is important to know that the kinds of pain we would want to alleviate determines the type of marijuana strain to be used. As a matter of fact, these diverse strains come with lots of in-depth qualities and properties that would suppress these chronic pains as they come. There are basically three types of cannabis strain that would relieve you of pain and they are:

1. Indica
2. Sativa
3. Hybrid

Be that as it may, there has been no tangible literature on the use of cannabis strain for alleviating pain or even other pain-related situations, whatsoever. Thus, I would be frank with you. Since the use of different strains in attacking different types of pain has not been generally accepted or even medically proven, using this would totally be unwise.

In the same vein, an online survey which was published in the Journal of Alternative and Complementary Medicine in 2014 was carried out and the results seem to be in favor of Indica strains for the alleviation of pain. And for boosting energy, the Sativa strain was highly preferred.

Additionally, the respondents leaned toward certain effects when using the cannabis plant as regards the alleviation of pain. To some extent, they agreed that the cannabis plant comes with effects like the neuropathy, spasticity, joint pain, and ordinary headaches. However the result might turn out to be, it still remains a survey. That is one of the limitations of it.

Additionally, it is an unethical and unprofessional survey whose result can be subjected to sentiments and outright guess. Thus, it shouldn't be taken too seriously. The respondents are not from the same place. Therefore, their answers are bound to be slightly different as regards drug use, dosage, and so much more.

With all we've pointed out, the studies, the research, and the points, the FDA still hasn't endorsed the use of cannabis for alleviating and suppressing chronic pain. According to the National Institute on Drug Abuse, the cannabis plant is not safe and iti would not be wise for one to follow the path. The results might be dangerous.

The FDA has not yet approved the use of any marijuana drugs for pain relief. However, the present situation of things is gradually pulling medical marijuana toward being the best alternative to suppress pain instead of the pain killer drugs out there. In the defense of this assumption, the marijuana plant usage in curing chronic pain comes with little or no side effects and there is no issue of overdose.

Research has also proven that cannabis plant intake as a direct remedy to chronic pain is considered safe and efficient. That way, the early concerns of the FDA are now being overlooked and the new face of medical marijuana now includes pain alleviation and suppression.

Additionally, the recent legalization of the cannabis plant has taken a new turn in the area of pain management and alleviation. Those patients that had been relying on the use of dangerous and harmful substances in the management of pain, whether it be acute, intense. or chronic pain, now changed sides to the use of the cannabis plant. This change in substance had been quite peaceful and harmonious - and the body generally adapts to the change with time.

Cannabis has proven to be a much better alternative and the safer way in soothing the pain away, no matter the level of pain you are feeling. No matter how painful it may feel or how intense it may turn out to be, making you extremely uncomfortable and inconvenienced, the best way to cure and suppress the pain is still via the cannabis plant.

It is the safest and healthiest way to approach pain. Pains like the toothache, sore muscles, ordinary and severe headache, muscle strain, general body pain, an injury or broken bone, dislocated joint, and so much more can be safely alleviated through the use of the cannabis plant. Narcotics and opiates might work too, but for how long and at what price?

Cannabis and chronic pain have always been grouped together. Right from the inception, it has been one of the easiest and safest means of alleviating and suppressing pain. However, it is your choice to use this medium or not. With that being said, let's delve into the next chapter, shall we?

Chapter Fourteen

Marijuana and Spirituality

The subject matter, "Marijuana and Spirituality" has been a trending combination in today's world with lots of advocates and supporters of the herb pointing toward the possibility of the cannabis plant being a spiritual symbol or something that can be used for something much more than smoking. They believe both concepts go hand in hand.

Ancient empires and kingdoms had earlier showed the truth and strength in the argument of these cannabis advocates with lots of archeological and literal evidence to prove their arguments right. However, all these points to the fact that the cannabis plant is definitely a supernatural plant.

Unlike the ordinary plants out there, the cannabis plant had earlier became the pathway to which the priests reach the spiritual realm in ancient times. Let's look briefly into the ancient spiritual use of the cannabis plant, shall we?

The old Indian Empire had been synonymous to the use of Bang in their worship. The Bang (cannabis) leaves are believed in their myths and legends to be created out of the body of their beloved Lord Shiva. According to the legend, Lord Shiva had created the plant from within himself, slept under the plant, and even ate it to refresh himself.

This is why priests during that era used the cannabis plant when indulging in spiritual activities. They would burn the leaves with some particular incense with the smoke filling the air and intoxicating them. This was believed to be the pathway of reaching lord Shiva. Also, during that era, the best gift you could ever bring as a sacrifice or alms is the cannabis plant.

Also, old Egypt is quite synonymous to the use of marijuana in their spiritual life. According to archeological and literally evidence, the Egyptians had attributed the plants to some of their gods. Thus, when worshipping them, the cannabis plant becomes a necessary ingredient. This was also extended to their ruling class.

That is how the cannabis plant can be spiritual. To this effect, you would agree with me that spirituality and the marijuana plant are two different but a connecting concepts that works closely together. In recent times, the plant has gone through a series of attacks and conviction by some groups of people who had and still believe that the plant has nothing beneficial to offer man.

Though, what had seemed like a forever struggle between both sides (those for and those against) can now be said to have reached its end as the people, states, and possibly countries are now seeing the light as regards this plant. They now see that there is no gain in criticizing the plant but will gain a lot in accepting it.

It would be beneficial to our economy in terms of jobs and income; it would be beneficial to our medicine in terms of finding a new cure; it would be beneficial to our lives and so much more. If we are now free to use and possess marijuana plant in some areas, then nothing should stop us from exploring the spiritual part of the plant. Nothing should stop us from reaching our inner self and chakra with the cannabis plant.

Additionally, using the cannabis plant any way we deem fit shouldn't be a problem, but where the problem should lie is when we start abusing and getting addicted to the plant. That alone is a big problem. With that, we tend to misuse the psychoactive and healing power of the plant. This doesn't harm the body in any way but there is really no point stuffing yourself with the plant. Even if you are taking it as a gateway from pain or stress, that is still not enough reason.

That is why the use of the cannabis plant as regards spirituality doesn't support this. In other words, the religions that view or see the cannabis plant as a tool or gateway towards reaching their spirituality do not accept this act of overdose. You can use it to ease or soothe your pain away, but never for the sake of being naughty.

Like we pointed out earlier, quite a number of people had been using this plant to reach their peak. It had been used in religious festivals, carnivals, events, and worship. For example, the Rastafarians make good use of the plant in their meditations and worship. They burn the leaves and gets intoxicated by the smoke. This would give them a clear head when reaching out to Jah (God).

They hold the belief that the cannabis plant was a gift given to them by Jah (God) in order to aid them in reflecting, thinking, and being insightful. According to them, they also believe that the Bible holds evidence that the cannabis plant is sacred and given to them by Jah (God) as a blessing. In Psalm 104:14;

"He causeth the grass to grow for the cattle and herb for the service of man...."

Also, in the book of Revelation 22:2;

"The herb is the healing of the nations."

They also hold the thought that the cannabis plant was in existence during the time of Moses. According to them, he had used the cannabis plant as a tool in reaching out to Jah (God). Aside from these Rastafarians, the Jews also argued that the herb can be seen in the Old Testament as it was mentioned more than once. However, when it was translated, the name was changed, thus, leading to the controversies out there.

Now, let me give a shocker! Do you know that a part of the Islamic sect is synonymous to the cannabis plant? Sufism, who is also known as "the hippies of Islam" had taken the cannabis plant mixed with a cocktail they call bhang. Traditionally, they believe the cannabis plant is the vehicle to spiritual ascending. In other words, the cannabis plant would open up their minds, relax their body, and pave way for divine intervention.

They believe the cannabis plant is not entirely banned or haram (forbidden), unlike alcohol that is outright forbidden. You can say the Sufis are more like the liberals of the religion of Islam. Thus, they don't interpret the Qur'an with great rigidity like the many other Islamic sects out there.

In addition to the Indian spiritual conscience of the cannabis plant, the cannabis plant is quite sacramental to the religions, most likely the Hindu. They see the cannabis plant as a sacred plant bestowed to them from their beloved Lord Shiva as a way of cleansing the world of its dirt and purifying one's soul. In other words, it is the elixir of life.

In Indian festivals like Holi, drinking bhang (also known as cannabis) is one of the important things that must be done during this festival. Afterward, there would be the distribution and spreading of colors to ward off evil and bring good. According to the ascetics, bhang is good for the body as it aids the mind from warding off during meditations.

The plant no doubt unifies one's mind, body, and soul. It also brings you closer to Lord Shiva. The sages also went ahead to attribute the functions and good luck of bhang to warding off negative repercussions and karma in the future. That way, one will only experience prosperity, good health, and plush with the cannabis plant. Lastly, making use of the plant recreationally or without any relation to religious activities is seen as a sin by these sages.

How to Use Cannabis Spiritually

How do you bring out the spiritual part of you with the cannabis plant? How do you attain and project spirituality while taking this plant? Instead of just smoking the plant for fun, how well can you get the best out of the plant? If you are looking for how to smoke this plant even better, then you should probably check another book. This chapter will help you reach your spiritual peak with the use of this sacred plant. Here are a few ways you can use to reach your spiritual peak with the cannabis plant.

1. Be aware of your tolerance level: It is important to know that the level of potency in the herb in recent times differs from that of the ancient days. You don't expect the recent techniques and skills added to the plant while growing it to have no effect on the plant. In fact, the taste, potency, scent, and many other qualities would definitely differ from that of the ancient cannabis. Thus, knowing the level of your tolerance for the plant is very necessary.

If it's too much, you can end up having a bad experience. This experience might not be as you had expected it to be. For example, smoking too much of the plant would simply do you no good. When you go above your limit, it will cause you to have a bad experience when trying to connect with your inner self. Intoxicating more smoke than you can handle will also leave you uncomfortable and inconvenienced in the end.

If this happens, then you know what to do. There is no need to force the connection. Pushing yourself is also not the way to go about it. Too much can cause you hallucinations and drowsiness. This is definitely not the best way to go as regards spirituality. Thus, when you want to begin this process, we would advise you get your dose right. Know how much cannabis you can take. Starting low is the right way to go. From there you can increase the dose with time.

2. Intentions must be made: There is nothing we would achieve without setting our intentions. No matter how little the project or activity might be, if we don't have the intentions to do it, we would definitely not do it. And if we end up doing it, we would surely not do it well. That way, intentions are everything. Be it a ritual or just meditation, having the intention is the first thing to do.

The connection between the cannabis plant and the intentions is vital. As the plant intoxicates you, your intentions will be your resolve into staying committed to the path that lies in front of you. No matter what your thoughts might be, be it soul searching, finding the inner self, reflecting on oneself, or absorbing the positive energy around you, intentions made will only help you keep focus.

Intentions will serve as a reminder toward staying in the game. That way, your eyes would stay on the prize and achieving the best out of the situation. There is no better way to attain complete focus.

3. Test yourself: After the first successful trial, there will surely be tons of questions going through our minds. We begin to ask ourselves questions like, was it actually worth it? Did we feel good afterward? Are we comfortable with it? Nevertheless, our mind would definitely be open to the possibilities around us. We would learn to accept ourselves and our environment even better.

And if it wasn't successful, we begin to ask ourselves why we even ventured into it. Our ego would definitely set in, opening doors for complaint. We would now start wondering if this psychoactive plant was even the best thing for us to begin with. Meanwhile, we would want you to embrace and learn from it. Use this as a milestone to get yourself in good shape.

This is the best way to test yourself. Walk through your ego and solve the mysteries surrounding your life. That way, you will see through the obstacles in front of you. You will only see life as simple as it can be. Trust me, testing yourself is the best way to see if the cannabis plant is working for you.

4. **Going for the right strain:** Knowing the right strain for you should be paramount as regards the cannabis plant. You can't just pick up a strain today and start getting intoxicated by it. What if the smell doesn't come out right? What if the potency is too strong for you to handle? Additionally, it is advisable for you to know these strains and their properties. Also, know the type of intoxication they give.

For example, the Sativas are known to be extremely rich in potency and they exude pure energy which increases thoughts. That way, they would aid the thinking faculty. The Indica species would also give a much better effect that would aid relaxation. It can also be sleep-inducing. To this effect, it is your choice to choose from any of these two or even more strains out there.

Be that as it may, choosing the right strain for you comes with the nature of your intentions, the extent to which your mental and physical energy can handle the potency of the plant, and even your level of determination. Don't hold back while selecting the best strain. You might end up not being able to make enough progress no matter how hard to focus or concentrate.

5. **Don't force it, let the plant just take control:** Even when everything doesn't add up, when you don't seem to feel your inner self, don't force your way in. Do not force the connection as you would only end up hurting yourself in the process. You have to trust the plant to take you through the journey swiftly and calmly. And if it doesn't suit your style and ways, then there really is no force in doing it.

Look for another alternative in reaching your inner self. Also, endeavor to show the cannabis plant the respect that it deserves. This respect comes in both ways. If you respect the plant while using it, the plant will also reciprocate by showing you the pathway to your journey.

We shouldn't forget that the cannabis plant contains more than just the THC. There are also lots of beneficial content that is present in the plant. These contents can aid your spiritual ascending. You would then find it very easy to reach the peak of your spirituality.

According to Bob Marley,

"When you smoke the herb, it reveals you to yourself."

This above quote puts the final nail on the coffin as the cannabis plant is bound to produce an intoxicating effect which would aid our journey toward being spiritual. If we manage to stay clean with the cannabis plant, we are sure to have improved health, good life, and surrounded with plush.

There has been quite a lot of research and studies on this field of the cannabis plant. Not very many people know about this aspect of the cannabis plant, but those aware of it are making sure they utilize every opportunity and positive vibe that comes from it. Like we said above, these benefits can only come as a result of your determination, intentions, and state of mind.

Some stoics even find themselves using this plant toward unlocking that indifference part of them. If the plant can clear one's mind, so can it create or form a certain mentality and psyche. We would now see the world differently. There would be a more profound effect that would enable them to see the world far from its material and worldly outlook. They would be able to focus more on the things that matter and not pain, pleasure, or emotions.

According to a study, the result shows that the cannabis plant has the power and necessary flavor to reactivate and re-energize a numb part of the brain, thereby leading the mind and soul to a whole new world. It will help rebuild and remove every possible obstacle that is standing in the way of your spirituality. In the end, your mind will be free and peace will engulf every part of your being.

Little wonder why smokers of the plant often don't see material things as something important. Most of them end up looking shabby, living in an ordinary building with little beautification and focus their time reconnecting with their true being. They tend to let go of any form of fear or anxiety and embrace the new world with peace, wisdom, and serenity. This is what the plant promises. All you need to do is allow it to take control.

Spiritual Effects of Marijuana

The spiritual effects of the cannabis plant can only be felt when we go mild on the intake. Most times, rushing the intake of this plant can be very disadvantageous as we end up having side effects. These side effects can come very strong at times. Thus, it is up to us to know the right amount of cannabis would be needed for an appropriate spiritual effect. Additionally, this spirituality comes in different forms - a long term experience and a short term one.

It all depends on how effective and efficient these strains of plants can be. Other factors that can influence this effect are the type of dosage taken, how deep your intentions can get, the physical and mental state of your being, and most especially the deep silence that is most needed.

Without this silence, there really can be no concentration. Imagine trying to focus in a room upstairs and your siblings are downstairs blazing the home theatres with loud music. Would you be able to focus in such environment? The answer is No. Thus, the mind needs a specific amount of silence in order to be able to connect with other realms.

Our minds will then be clear and our thoughts sieved to meet the requirements of getting a true connection. Our brain would shut out unnecessary and irrelevant thoughts, thereby, focusing on what truly matters. Your illusions would be transformed and your mind would serve as the window to the other realm. An amazing level of relaxation and calmness would naturally come after this exercise.

As a user of the plant, you will be forced to let go of negative thoughts as they would serve as an obstacle toward your spiritual growth. Negative thoughts will always serve as a hefty load which will weigh on your mind. They would make the mind bulky and stuffed. What the cannabis plant would do is to make sure that the mind unhooks and unbuttons these thoughts and drop them in order to be free. In the same vein, you will no longer feel stressed and calmness will always reign over you.

Quite a lot of people in the world today had taken the cannabis plant to be more than just a plant. It has been a source of inspiration and creativity to lots of people, especially artists, musicians, business innovators, and so much more. The plant dishes out stability and a clear head which would ensure that one's full potential is reached. It would help reduce tension and focuses our mind on the main goal.

In the same vein, the cannabis plant also shields the body from falling into the depth of emotional traumas and conditions. How does it exert this function? It simply breaks one's resolve by penetrating through our defenses. The cannabis plant would ensure these emotional resistances are broken, thereby building a much better condition for the person. Instead of holding up and suffering in silence, the cannabis plant would give you the strength to accept the situations as they come.

That is the only way you would truly be happy. Sometimes, it's really not about putting up a fight. Sometimes it's about accepting defeat and learning from your flaws. That way, you would be able to move on without having regrets. A life filled with regret is a very sad life. That is when you will start appreciating life and seeing the beauty of everything around you.

The road toward this level of spirituality has been in existence for thousands of years. Marijuana has been used in reaching greater heights as regards spiritual awakening. In fact, many people have sought out spiritual healing with the help of this plant and have since then gotten exactly what they had hoped for.

Like many other plants out there, the cannabis plant represents quite a number of things. It is not just the medical plant as it is popularly known in today's world, As a matter of fact, it has quite a lot of meaning and functions which are still being discovered to date. Some of its importance and functions are as follows, releases stress, emotional resistance, letting go, oneness, surrender, indifference, and so much more.

It is important to know that our level of response to the use of the cannabis plant in seeking spiritual healing or ascending can be quite different. While it might work for others that are highly dependent on the plant, it might not even work well for some people. Not that the cannabis plant does not really work for everyone that is willing to really try it, but what it might need is just a little bit of adjustments and tweaking.

That should do the trick. What do you want? What does your body react to positively? How much cannabis can you take and what strain do you feel is perfect for you? These and more are the questions you need to ask before jumping into this adventure. However, there have been critics of this line of spirituality and they have been gaining momentum in recent times.

What if one becomes an addict on the verge of reaching their spiritual peak? This is very possible as regards using marijuana for spiritual purposes. If you feel you would like to try out a cannabis plant strain while becoming a guru in your spiritual expedition and end up becoming a chronic addict to the plant, how would you solve that? What would you do?

A lot of people have fallen into this trap without finding a successful solution to their addiction problem. Some of them had even tried taking the cannabis plant in breaking off their emotional resistance, so as to rebuild themselves in a better light. But the result had not been quite as pleasant as they had expected it to be.

They may have tested themselves too high or set the limits too deep for them to pull out. This brings us to the question of how can we measure these limits. How can we know when is the best time to pull ourselves out? What if it ends up overpowering us and making us stayed buried in the experience? The best way to stay sober would now be to keep taking the cannabis plant.

While the cannabis plant might offer a very calm and easy way to reach out to our illusions and connect with the other realm, it could also mean a complete trap which would eat us up. One minute it would place up pictures of what we really want and the next, it will just snap them hard before we could even respond. This is what the critics of marijuana spirituality are saying.

In their submission, they are saying the cannabis plant works for spiritual ascending perfectly well, but they aren't the best process to follow. As a matter of fact, they are not even necessary for this lime of spirituality. They believe there are other steps to follow in order to reach the inner self and connect with the other realm.

Spirituality or not, the cannabis plant is indeed a blessing to the world. When the old Indians called it the elixir of life, there weren't particularly wrong with the diverse functions and multidimensional properties that can be found in the plant.

Be that as it may, the plant is great for spiritual purposes. Inasmuch as it may have its own flaws as regards spirituality, it is still one of the best vehicles for spiritual ascending. Get a cool, quiet place, sit or lie in an appropriate manner, burn the cannabis plant, inhale, and let the plant take you on the journey of knowing your true self. The last chapter will give you a quick and short recap of the book. In case you still haven't gotten anything in the book, here is your last chance.

Chapter Fifteen

Now You Know Better!

Growing marijuana is not as easy as it sounds. It should take more than just a few months to harvest and start enjoying the fruit of your labor. Growing the plant, as we have explained in the previous chapters takes lots of patience, understanding, and commitment. From the problem of pests (human and natural) to the problem of security, finance, location, soil, and so much more.

Maintaining the plants is even harder. This is when you would start monitoring and paying close attention to the plants all through the phases we had earlier mentioned. One slight negligence could mean the death of your whole garden. For example, leaving an infected strain within the midst of others would definitely affect the rests with the infection, thereby killing off the plants.

When this happens, uproot the plant and if possible and change the soil so as to be on the safer side. You can always refer to this book in case you are lost or don't know what to do. After reading and fully understanding this book, no one should be able to tell you a lie concerning the cannabis plant. To even make fun out of the situation, you can even pretend as if you don't know anything concerning the cannabis plant.

It once happened to me when I was going to pick up my daughter in school. As a quiet and easy-going fellow, I had not always seen the need to interact with either the teachers or fellow parents each time our paths crossed. A simple gesture with the hand does the trick.

This changed the moment I schooled one of her teachers on the positive side of the plant. He was discussing with a fellow teacher on the negative side of cannabis. He seemed to be against the legalization of the cannabis plant as well as churning up things to crucify the plant. If I hadn't known better, I would say he is an ignorant fellow. But he is a teacher, my kid's teacher for that matter. So I chipped into their conversation calmly and maturely.

I just knew I had to say something, else I won't ever forgive myself for turning a blind eye. So I took my time to explain things to him gradually. I started from the historical background and benefits the plants had given to the ancient empires and kingdoms before ending the lecture on the recent breakthroughs in medicine, all thanks to medical marijuana.

Ever since, the teachers have been looking at me with respect and appreciation. I had known better and that is why I made sure to share the little knowledge I had on the plant. Why not do the same when the need arises and the situation warrants you to? There is an amazing feeling that comes with such enlightenment.

Be that as it may, let's do a quick recap on the whole book, shall we?

The book has delved into the steps, ways, and methods of growing cannabis. Chapter one of the book focused on what cannabis really entails. It went further to explain and shed more light on the historical background of the cannabis plant. How did this plant come into being? It's a historical background with case studies (the Chinese empire, the Egyptian empire, the Greek-Roman empire, and so much more). Did these diverse kingdoms really make use of the cannabis plant? How much and how well did they make use of it? This chapter answers all these questions and more.

Chapter two of the book focused on the face of marijuana in today's world. Over the years, the battle and struggle for legalization the plant had been quite stiff. To be frank, it had been an off and on business. The 19th century saw the wide acceptance of the plant for both recreational and medical purposes. While the 20th century saw the struggle for legalization and stereotypes.

The 21st century is seeing a new trend of acceptance in the line of marijuana through medical cannabis. It had become more than just a plant. It had opened up the eyes of many as regards its benefits.

Chapter three is all about reestablishing the fact that cannabis addiction is misplaced and misconstrued. There is no such thing as cannabis addiction, or should I say the term addiction is being misconceived. This chapter would shed more light on this subject matter.

Chapter four talks about cannabis horticulture. Horticulture entails the agricultural processes of the cannabis plant, the process of growing the plant, the techniques, the making of hashish, and many other concepts that involve marijuana. This chapter will open up your eyes to the growing of the plant.

Chapter five delved into the two methods of growing cannabis – the indoor and outdoor methods. The chapter will further talk about how to go about preparing and putting your plans to grow cannabis into action. The chapter will show you how to grow the plant indoors and outdoors.

Chapter six is all about the three most important stages of the plant. While growing the cannabis plant, there are certain care and attention we must give to the plant. This chapter will give you the knowledge to tackle any problems as they come.

Chapter seven delved into the harvest of this plant. After growing the plant, maintaining it, and following every step toward achieving a healthy plant, the next thing to do is to reap the fruit of your labor. There are similar but distinct ways of harvesting both indoors and outdoors cannabis. This chapter will show you how to harvest them both.

Chapter eight talks about cannabis businesses and the present face of the industry. It would interest you to know that the cannabis industry is very much alive and is contributing its own quota in the American economy. The chapter will familiarize you with everything you need to know about the industry.

Chapter nine is about the steps to follow toward becoming a topnotch businessman or woman in this field. Starting as a novice doesn't necessarily mean you can't develop into being a giant in this line of business. Follow these steps carefully and closely and the rest would be completely easy.

As an investor, I would want you to focus more on chapter ten. It will open up your eyes on this kind of business with the opportunities surrounding it. As a growing industry, it would be wise as an investor to secure your place in the industry and write your name in ink. Don't forget to read through the risks involved as an investor.

Chapter eleven is all about traditional medical marijuana, from the historical background to the medical benefits it gave the old empires and kingdoms. Everything is being captured in this chapter. Now the question you should ask to yourself after reading this chapter is if the old kingdoms are all about the cannabis plant and its medicinal properties, why are we then against the plant?

Chapter twelve looked at the struggles and glories of the cannabis plant as regards medicine. The chapter will show you how modern-day medical marijuana is the new face of medicine. This will bring us to the next chapter. In this chapter, the cannabis plant can be used as an alleviator and reliever of pain. Chronic pain can be suppressed with the use of the cannabis plant. Pain is the alarm of our body system.

Chapter fourteen makes it the semi-final chapter. It delved into the relationship between the cannabis plant and Spirituality. This chapter looks at how the cannabis plant serves as a vehicle for spiritual ascending and awareness. That should put the lid to our bottle of knowledge as regards this book.

Now you know better. Now you are more informed. Now you are more knowledgeable as regards the cannabis plant. It's left for you to either make use of it or keep it to yourself. The choice is entirely yours.

Conclusion

No doubt, it had been an exciting experience since the beginning of this book. Each chapter of this book will open up your eyes to the positive side of the plant. At the end of this book, one thing should be made clear to you and that is the new knowledge you've gotten from the book. Obviously, your life will not remain the same as you are now familiar with lots of concepts, terms, and keywords related to the marijuana plant.

From growing marijuana to starting a career in the cannabis industry, this book will walk you through every step of the way. After knowing all that, it is left for you to either capitalize on the new experience or, better still, share the knowledge to the best of your abilities.

Instead of going through vigorous exercises and activities in attaining spiritual ascending, the cannabis plant would aid you in achieving this state. According to a popular saying, to know is to forget and to practice is to remember. To this effect, don't just say you know how to go about pulling off a cannabis garden, starting a Canna-business, using cannabis as an alleviator for pain or even going spiritual with the plant.

Be practical with it. See if you really understand the whole concept of the book. You can't tell if you will need to read the book all over again. And in cases where you miss some steps or get lost on when growing the plant, the book is always there to guide you.

When people who have been in the cannabis business for long start seeing the different strides you are making in this line of business, even as a novice, they won't know it's because you have devoted yourself to our book. Also, if as a young investor you already know the risks and strengths of the marijuana industry, it's very normal for people to start getting amazed when you start swerving through the obstacles in the industry. They won't know it's because of this book.

Go out there, make your mark, and don't forget to write your name in paint. If you are scared or having cold feet before, don't be now. This is a new dawn for marijuana. Right now, more than half of the states in the country have laid down their guard as regards the cannabis plant. And according to forecasters, many more will follow suit in due course as time passes. Thus, if you are still in a state where it is prohibited, then take a chill.

In the end, you are going to be making a lot of money in this industry. Be that as it may, I would like to draw the curtains with this quote:

"The marijuana plant is one of the biggest blessings mankind has ever known. Due to obvious reasons, the innocent plant was being stereotyped and criminalized. But these accusations aren't true. Stay strong to your beliefs. Don't let anyone sway you from them."

Thank you for being patient with me during the course of this mind-blowing experience. God bless and bye for now!

www.ingramcontent.com/pod-product-compliance
Lightning Source LLC
Chambersburg PA
CBHW071436080526
44587CB00014B/1878